SpringerBriefs in Plant Science

More information about this series at http://www.springer.com/series/10080

Girdhar K. Pandey · Poonam Kanwar
Amita Pandey

Global Comparative Analysis of CBL–CIPK Gene Families in Plants

 Springer

Girdhar K. Pandey
Poonam Kanwar
Amita Pandey
Department of Plant Molecular Biology
Delhi University South Campus
New Delhi
India

ISSN 2192-1229   ISSN 2192-1210 (electronic)
ISBN 978-3-319-09077-1 ISBN 978-3-319-09078-8 (eBook)
DOI 10.1007/978-3-319-09078-8

Library of Congress Control Number: 2014944548

Springer Cham Heidelberg New York Dordrecht London

Printed on acid-free paper

Springer is part of Springer Science+Business Media (www.springer.com)

Preface

Ever since agricultural cultivation came into existence, plant systems and their functions are manipulated rapidly both by humans and nature. Biotic and abiotic stresses adversely affect crop productivity. Plant breeding and genetic engineering have enabled maximizing crop productivity even under minimal stress conditions. However, such tools also have their limitations and so the reverse genetic and functional genomic approaches came into practice. With the availability of genome sequences in several plant species, the major focus has been on identification of the functional role of several signaling networks and components involved in regulating stress tolerance pathways. A signaling cascade begins with sensing the stimulus, and the basic outcome of any signaling machinery is to bring out the final response. But at the start and end of any phenomenon, there are certain characters that play their role to assist the process. There are various molecules inside the cellular machinery, which are involved in transducing the signals.

Calcium plays a pivotal role in regulating the physiological and developmental processes in plants. Till now, several calcium sensors have been discovered, which regulate the diverse signaling pathways involved in plant growth and development. One of the major calcium sensors CBL (calcineurin B-like) is decoding the calcium signal during various environmental stresses in plants. Calcium-mediated signal is transduced downstream by CBL-interacting protein kinases (CIPKs), which generally phosphorylate the target proteins such as transcription factors or transporters/channel leading to a response. Mutant-based approach has provided valuable information in the functional analysis of individual members of *CBL* and *CIPK* gene family in Arabidopsis. Both *CBL* and *CIPK* gene families have previously been identified and characterized in Arabidopsis and rice. Identification and characterization of *CBLs* and *CIPKs* in other plant species such as *Oryza sativa*, *Pisum sativum*, *Cicer arietinum*, *Zea mays*, *Populus euphratica*, *Vitis vinifera*, *Malus domestica*, *Gossypium hirsutum*, *Sorghum bicolor*, *Brassica napus*, *Vicia faba*, *Phaseolus vulgaris*, *Ammopiptanthus mongolicus* and *Triticum aestivum* are still at a juvenile stage.

An up-to-date research in the field of CBL–CIPK signaling transduction components, basic and advanced, in various genera has been reviewed here. Also, current understanding and knowledge of genome-wide organization, evolution, structure, sub-cellular localization, and biochemical properties of CBLs and CIPKs in plants is discussed in various sections of this book. In the first few sections, there is a basic overview of the field of CBL–CIPK signaling, which is accompanied by genomic organization of CBL and CIPK component throughout the plant kingdom. CBL and CIPK proteins distributed in the plant cell and also in different tissues and organs might be involved in executing a particular function. The expression analysis of *CBLs* and *CIPKs* during different developmental stages as well as under different physiological conditions also hints toward their functional role during these particular conditions. So, not only the change in mRNA level but protein modification may also be a strategy to provide specificity for CBL–CIPK signaling. In the subsequent sections, phosphorylation, interaction, and biochemical properties of CBLs and CIPKs are discussed. Finally in a few sections, the targets of CBL–CIPKs and the functional role identified for several CBLs and CIPKs during different environmental stresses, nutrient deficiency, and developmental stages is discussed case-by-case in Arabidopsis and other higher plant species.

Overall, this comprehensive study is focused on the diverse role of the CBL–CIPK module in different stress signaling and also to identify a newly emerging role of this calcium-signaling module in plant growth and development across different plant species. In addition, besides Arabidopsis, it provides a backbone of knowledge to perform detailed molecular investigation in crop plant species and could possibly enable in designing strategies to tame abiotic stress tolerance and development in important agronomical crop plants. This book will act as a handy and informative source in this field for students as well as advanced researchers.

Acknowledgments

We are thankful to the Department of Science and Technology (DST) and the Department of Biotechnology (DBT), India, for supporting the research work in GKP's lab. We also express our thanks to Dr. M. C. Tyagi, Mr. Sibaji Sanyal, and Mr. Shashank Maurya for critical reading of this manuscript.

Contents

1 Basic Terms and Overview of Contents................................ 1
   1.1   Introduction 1
   1.2   Calcium on the Way of Signaling....................... 2
   1.3   CBLs: The Calcium Sensor 2
   1.4   History and Concepts................................. 4
         1.4.1   Discovery 4
         1.4.2   New Paradigm 6
   References.. 7

2 Genomic Organization 13
   2.1   Introduction 13
   2.2   CBL and CIPK Complements 15
         2.2.1   CBL- and CIPK-Type Proteins in Protozoan.......... 15
         2.2.2   CBL and CIPK in Algae........................ 15
         2.2.3   CBL and CIPK in Higher Plants 15
   2.3   Genomic Architecture 16
   2.4   Gene Structure 16
   2.5   Phylogenetic Relatedness and Evolution 17
   References.. 17

3 Distribution and Expression in Plants........................ 19
   3.1   Introduction 19
   3.2   Distribution of CBLs and CIPKs in Plants 20
   3.3   Expressions under Various Environmental
         and Developmental Conditions 21
   3.4   Expression of Stress Markers Genes in Mutant
         and Overexpressor of CBL and CIPKs 23
   References.. 25

4   Protein Structure and Localization 29
   4.1   Introduction 29
   4.2   Motifs and Domains................................... 30
          4.2.1   Motifs in the CBLs............................ 30
          4.2.2   Motifs in the CIPKs 31
   4.3   Protein Structure..................................... 31
          4.3.1   Three-Dimensional Structure of CBLs 31
          4.3.2   CBL–CIPK Complexes 32
   4.4   Subcellular Localization................................ 32
          4.4.1   Subcellular Localization of CBL Gene Family 32
          4.4.2   Subcellular Localization of CIPK Gene Family 33
          4.4.3   Subcellular Targeting of CBL–CIPK Complexes 33
   References.. 35

5   Biochemical Properties of CBLs and CIPKs 39
   5.1   Introduction 39
   5.2   Mutagenesis of CIPKs to Generate Hyperactive
          Kinase or Dead Kinase................................. 40
   5.3   Function of Auto-phosphorylation in CIPKs 41
   5.4   Physiological Target/Substrate of CIPKs................... 41
   5.5   Phosphorylation of CBL by Their Interacting CIPK 42
   References.. 43

6   Protein Interactions 45
   6.1   Introduction 45
   6.2   Various Interactors of CBLs............................ 46
          6.2.1   CIPKs..................................... 46
          6.2.2   Others..................................... 46
   6.3   Various Targets of CIPKs.............................. 46
          6.3.1   Phosphatases 47
          6.3.2   Transporters/Channels 47
          6.3.3   Transcription Factors 48
          6.3.4   Enzymes.................................... 48
   6.4   CBL–CIPK Complexes Regulate a Broad Range of Functions... 49
   References.. 49

7   Functional Role of CBL–CIPK in Nutrient Deficiency 51
   7.1   Introduction 51
   7.2   Nitrate Deficiency 51
   7.3   Potassium Deficiency.................................. 53
          7.3.1   CBL–CIPK23–AKT1........................... 53
          7.3.2   CBL2/3–CIPK9-Unknown Target 55
          7.3.3   Other CBL–CIPKs Regulating AKT1 56

	7.3.4	CBL4–CIPK6–AKT2.	58
	7.3.5	CBL–CIPK Regulating K+ Nutrition in Other Plants Species.	59
	References.	60

8   Functional Role of CBL–CIPK in Abiotic Stresses 65
  8.1 Introduction 65
  8.2 Salt Stress. 66
  8.3 Drought and Osmotic Stresses 68
  8.4 Cold Stress 69
  8.5 ABA Signaling. 69
  8.6 pH Stress 72
  8.7 Flooding Stress. 73
  References. 74

9   Functional Role of CBL–CIPK in Biotic Stress and ROS Signaling 79
  9.1 Introduction 79
  9.2 Biotic Stress and ROS Signaling 80
  References. 81

10   Functional Role of CBL–CIPK in Plant Development 83
  10.1 Introduction 83
  10.2 Pollen Germination and Tube Growth 83
  10.3 Flower Development 84
  10.4 Root Development 85
  10.5 Seedling Development 85
  References. 85

11   Application and Future Perspectives of the CBL–CIPK Signaling. 87
  11.1 Basic Study Done So Far 87
  11.2 Applications of the CBL–CIPK Signaling System 88
  11.3 Future of CBL–CIPK Signaling. 89
  11.4 Questions for the Future. 90
  References. 90

Chapter 1
Basic Terms and Overview of Contents

Abstract Calcium plays a pivotal role in regulating the physiological as well as developmental processes in plants. Till now, there are several calcium sensors discovered, which regulate the diverse signaling pathways involved in plant growth and development. One of the major calcium sensors is calcineurin B-like (CBL) decoding the calcium signal during various environmental and physiological processes in plants. Calcium-mediated signal is transduced downstream by CBL-interacting protein kinases (CIPKs), generally phosphorylating the target proteins such as transcription factors or transporters/channel and finally leads to generation of a response. A signaling cascade possesses defined set of CBL–CIPK proteins and several designated target proteins.

Keywords Calcium · Signaling · CBL · CIPK · Sensor · Discovery · New paradigm · Calcineurin · Genome-wide

1.1 Introduction

Among multi-cellular organism, higher plants and animals have developed complexity in terms of growth and development. Even though both are fundamentally different from each other, plants have developed more intricate mechanisms to sense and react to the environmental changes during the course of evolution, majorly due to their sessile nature [3, 50]. In the given environment, plants constantly encounter a plethora of favorable and unfavorable stimuli. Unfavorable stimuli encompass diverse environmental changes such as nutrient deficiency in the soil and various abiotic and biotic factors, leading to stressful conditions for normal growth and development of plants.

To cope up with this, plants have developed complex and intricate signaling networks, which enable them to develop adaptive responses against these stimuli. Signal transduction pathway in plants comprised of second messengers, sensors, and effector proteins. One of such second messengers is calcium, which perceives the signal from membrane receptors and switch on intracellular signaling cascade (Fig. 1.1).

© The Author(s) 2014
G.K. Pandey et al., *Global Comparative Analysis of CBL–CIPK Gene Families in Plants*, SpringerBriefs in Plant Science, DOI 10.1007/978-3-319-09078-8_1

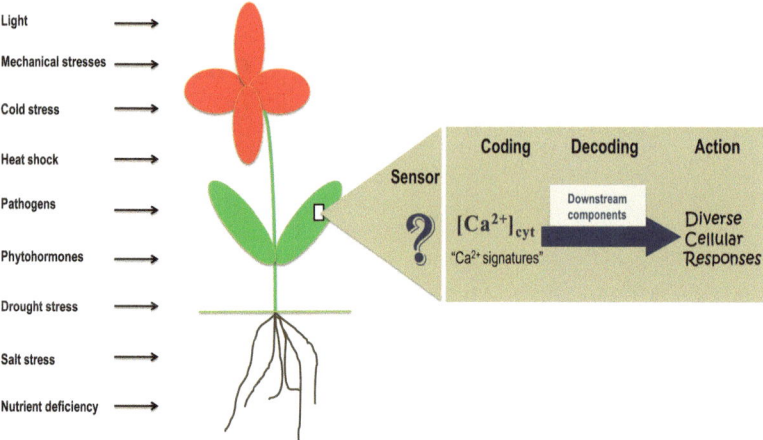

Fig. 1.1 Role of calcium signaling in managing environmental stresses. To cope up different kinds of environment changes, several complex sets of sensing and signaling machinery are activated in the plant cell. Calcium signaling is one of the complex and intricate signaling networks, enabling plant to develop adaptive responses

1.2 Calcium on the Way of Signaling

Calcium as a secondary messenger works by generating spatially and temporally unique Ca^{2+} oscillations inside the cell called 'Ca^{2+} signatures' for a specific response [33, 42]. Coordinated activities of Ca^{2+} permeable channels/pumps and active transporters located in either at plasma membrane or at different endomembrane are responsible for such specific 'Ca^{2+} signatures' [32, 54].

Further in the signaling pathway, intracellular increase in Ca^{2+} is perceived by various Ca^{2+}-binding proteins, Ca^{2+} sensors, such as calmodulin (CaM) [42], Ca^{2+}-dependent protein kinases (CDPKs) [6, 23, 46, 52, 57, 75], calcineurin B-like proteins (CBLs) [3, 41, 42, 47], or other Ca^{2+}-binding proteins [51, 56] to initiate various cellular signal transduction events (Fig. 1.2). Activation generally leads to changes in conformation of the calcium sensors, such as CaM and CBLs [42, 54, 77]. Calcium signaling involves several protein kinases, phosphatases, and their various targets that transduce the Ca^{2+} signal further downstream [15, 16]. Figure 1.2 shows the event diagrammatically.

1.3 CBLs: The Calcium Sensor

CBLs are Ca^{2+} sensor in the plant cells, which regulate the activity of CBL-interacting protein kinase (CIPK) (Fig. 1.3).

CBL–CIPK networking is involved in diverse number of signaling pathway [1, 3, 42, 45, 58]. Members of the CBL and CIPK family are found exclusively in

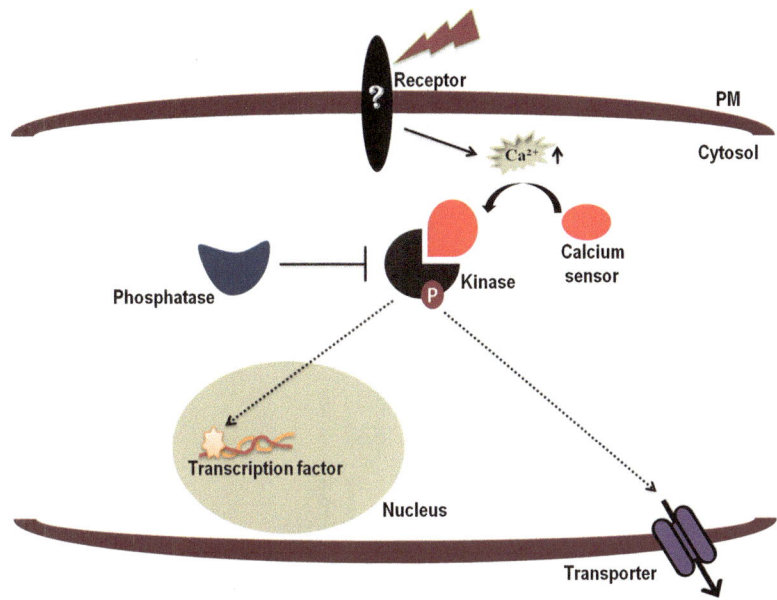

Fig. 1.2 Hypothetical model for a signaling cascade activated by calcium. A stimulus is sensed by receptor/sensor at the plasma membrane, which might lead to increase of calcium concentration in cytosol, a primary event in this response. Calcium signaling involves various calcium sensors, protein kinases, phosphatases, and their various targets that transduce the Ca^{2+} signal downstream in signaling pathway. Here, in this signaling cascade, a transcription factor and transporter shown might act as possible targets of kinase

the plant kingdom. 10 CBL and 26 CIPK genes are identified in Arabidopsis [3, 20]. Exploration of these gene families in Arabidopsis has suggested their role in various environmental stresses. The close pairing of similar isoforms in rice and Arabidopsis suggests potential orthologous relationships for CBLs and CIPKs [34]. Therefore, exploration of CBL–CIPK networking is becoming more important in the other agronomical important plants for developing a better understanding of calcium-mediated signaling in plants.

There are recent reports on CBLs and CIPKs in various economically important crops, cereals: *Oryza sativa* [36, 37, 49, 66–68], *Zea mays* [8, 26, 60, 64, 76], *Triticum aestivum* [11], *Hordeum brevisubulatum* [38]; legumes: *Pisum sativum* [45, 63], *Cicer arietinum* [62], *Vicia faba* [61], *Phaseolus vulgaris* [19], and *Vigna unguiculata* [24]; fruits: *Malus domestica* [22, 65], *Solanum lycopersicum* [70], and *Vitis vinifera* [9, 10]; fibre plants: *Gossypium hirsutum* [14]; oilseed plants: *Brassica napus* [7]; woody plants: *Populus euphatica* [71, 73] and *Populus trichocarpa* [69, 72]; and plant like *Ammopiptanthus mongolicus* [8, 17, 26, 55], which are agronomically important species vulnerable to both biotic and abiotic stresses and require detailed understanding of signal transduction mechanism to generate stress tolerant varieties. A detail genome-wide analysis of several gene families in these plants has greatly helped to advance our understanding of

Fig. 1.3 CBLs are the calcium sensors, which interact with kinases to transduce the signal downstream in the signaling pathway. Calcium sensors might act as sensor responder or sensor relay to transduce the signal downstream in a typical calcium signaling pathway. CBLs relay signal by interacting with CIPKs, downstream in the signaling pathway

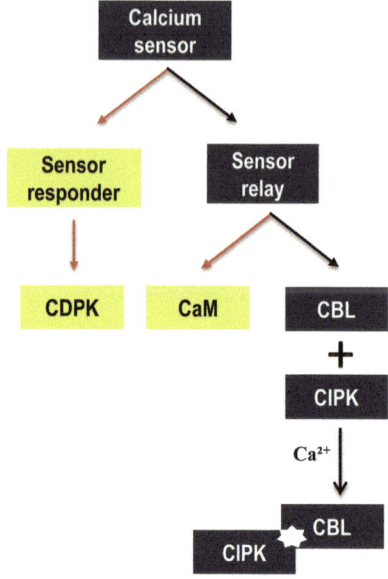

stress-mediated signaling. By investigating the molecular mechanism of calcium-mediated CBL–CIPK signaling pathways in model plants Arabidopsis and rice, we can extrapolate the information to other crop plants, to tackle the problem of crop loss due to abiotic and biotic stresses.

1.4 History and Concepts

1.4.1 Discovery

Calcineurin is a Ca^{2+} and calmodulin-dependent serine/threonine phosphatase [30, 31, 59]. It is heterodimeric protein consisting two subunits. Catalytic subunit called (CNA), which interacts with calmodulin in a Ca^{2+}-dependent fashion and a regulatory (CNB) subunit, composed of helix–loop–helix structural motif (the EF-hands) [31]. B subunit of calcineurin interacts with calcium, crucial for binding to substrate. In animals and yeast, calcineurin regulates various physiological and developmental processes [53]. Several reports suggested the presence of calcineurin-like activity based on biochemical characterization in few plants [2, 43, 48]. However, the genes encoding the two subunits of calcineurin, i.e., CNA and CNB, were not identified even after the post-genomic era in several plant species including Arabidopsis and rice. Moreover, in an extensive hunt for identification of calcineurin or related protein from plants, two major groups identified similar protein known as calcineurin B-like protein or CBL by genetic as well as biochemical studies [35, 40]. By using genetic screen, Zhu and co-worker [40] identified salt overly sensitive (sos3) mutant, a calcium-binding protein. At the same time,

Fig. 1.4 A parallel calcium signaling pathway in animals/fungi and plants. Calcineurin B-like protein (CBL), a Ca^{2+}-binding protein of Arabidopsis similar to regulatory B subunit of calcineurin (CNB) of animals, associate with CIPK to transduce the signal downstream

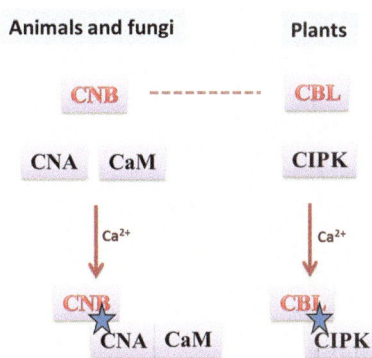

by using the rat CNA as bait in yeast two-hybrid screening of plant cDNA library, Luan and colleagues [35] identified similar calcium-binding protein called CBL, sharing high homology with calcineurin-B subunit (CNB) of yeast and neuronal calcium sensor (NCS) from animal [35] (Fig. 1.4). CBLs contain EF-hands such as CaM, NCS, and CNB [34, 42].

Following the discovery of CBL in Arabidopsis, a specific group of protein kinases were identified to be associated with CBLs in calcium-dependent manner [18, 27, 58], sharing similarity to SnRK3 subgroup of plant kinases [21]. These findings link a new type of Ca^{2+} sensors to a group of novel protein kinases, providing the molecular basis for unique Ca^{2+} signaling machinery in plant cell. A 24 amino acid domain, NAF domain, unique to CIPKs was identified as being required and sufficient for interaction with all known CBLs [1]. First, rice ortholog was identified by Kim et al. [28], when a cold-inducible gene (designated *OsCK1*) from *Oryza sativa* shared similarity with CBL-interacting protein kinases (CIPKs) of Arabidopsis. In conclusion, the discovery of CBL–CIPK proteins in plants led to development of a new paradigm in calcium signaling.

In yeast and animals, calcineurin is comprised of two subunits, a PP2B phosphatase (CNA) and a calcium sensor subunit (CNB). However, in case of plants, the PP2B phosphatase, i.e., CNA is not identified yet, and in place of CNB, plants do possess a closely related protein called CBL, which interact with a specific group of kinases called CIPK (CBL-interacting protein kinase). As depicted in the Fig. 1.4, plants have acquired a novel calcium signaling components involving a calcium sensor and kinase module instead of two calcium sensors, i.e., CNB and CaM, which activate PP2B-type phosphatase, called calcineurin-A subunit (CNA).

After the availability of genome sequences of Arabidopsis and rice, Kolukisaoglu et al. [34] presented a comparative genomics analysis of full complement of CBLs and CIPKs in Arabidopsis and rice. Ten members of CBLs were identified in both Arabidopsis and rice. High degree of sequence similarity was found in both the species [34]. Twenty-five CIPKs were identified in Arabidopsis, whereas in rice, a complement of 30 CIPKs was reported initially [34, 42]. One more Arabidopsis CIPK member was identified later, i.e., *CIPK26* [12, 29, 44]. Similarly, a few other members were also found in the rice genome (*OsCIPK31*,

Fig. 1.5 The emerging functional role of CBL–CIPK networking. Role of CBL–CIPK was earlier known only in environmental or abiotic stresses but recent work in CBL–CIPKs signaling in Arabidopsis and other plants speculated their involvement in development, biotic stress, and ROS signaling

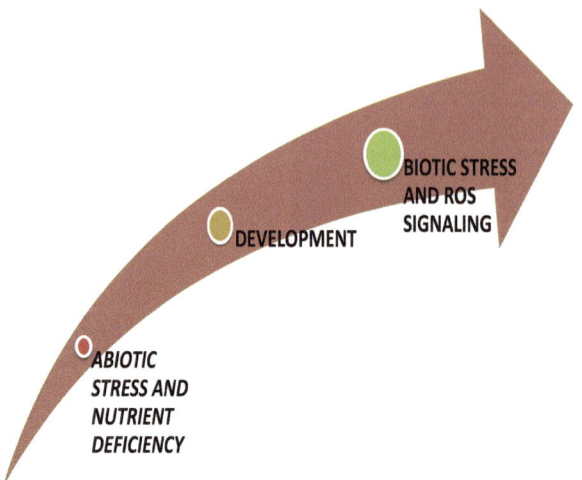

OsCIPK32, and *OsCIPK33*) [49, 74]. From the comparative analysis of CBLs and CIPKs calcium signaling network, a similar structural and functional role was speculated in both the plant species.

Because of complete annotated genome sequence and availability of large number of molecular tools in Arabidopsis, several publications specifically in Arabidopsis have emerged compared to other plant species. Various reports on genome-wide and functional characterization of CBLs and CIPKs in different plant species have also been reported over a period of time. With the data obtained from other plant species related to *CBL* and *CIPK* genes, a high level of conservation has been observed structurally and functionally.

1.4.2 New Paradigm

Recent reports on the regulatory aspect of CBL–CIPK pathway enabled to recognize novel dimension of this signaling network in biological system. One of the most important and fascinating facts about CIPKs is that they not only phosphorylate the downstream target but can also phosphorylate CBLs [13, 20, 39, 45]. Phosphorylation of Arabidopsis CBL occurs at a serine residue within a conserved C-terminal "PFPF" motif [13], affecting stability of CBL–CIPK complex and is important for target regulation [13, 20]. CBLs can be post-translationally modified by lipid modification such as myristoylation and acylation or palmitoylation, which imparts them another important biological regulatory property by targeting them to plasma membrane [4, 25]. N-termini of some of the CBL members harbor an MGXXX(S/T) motif that allows attachment of myristic acid to the glycine residue [4, 25]. In addition to myristoylation, CBLs can be N-terminally S-acylated/palmitoylated in order to be functionally active to target the CBL–CIPK complex to plasma membrane [4]. The differential subcellular targeting of CBL–CIPK complexes is a mechanism for transmission of spatially defined signals to different destination in the cell. These results among others firmly established that CBL proteins determine the localization and site of action of CBL–CIPK complexes [4, 5].

Since the discovery of CBL–CIPKs, most of the CBLs and CIPKs are well characterized in model plant Arabidopsis in abiotic stress signaling. Some of the CBL–CIPKs are also speculated to be involved in biotic and/or oxidative stress and developmental pathways (Fig. 1.5). Extensive studies reporting the functional role of CBLs and CIPKs in plant species besides Arabidopsis will greatly advance our understanding of their diverse functional role in regulating different biological processes and develop some useful tools to genetically modify the crop plants to enhance the quality and quantity.

References

1. Albrecht V, Ritz O, Linder S, Harter K, Kudla J (2001) The NAF domain defines a novel protein–protein interaction module conserved in Ca^{2+}-regulated kinases. EMBO J 20:1051–1063
2. Allen GJ, Sanders D (1995) Calcineurin, a type 2B protein phosphatase, modulates the Ca^{2+}-permeable slow vacuolar ion channel of stomatal guard cells. Plant Cell 7:1473–1483
3. Batistic O, Kudla J (2004) Integration and channeling of calcium signaling through the CBL calcium sensor/CIPK protein kinase network. Planta 219:915–924
4. Batistic O, Sorek N, Schultke S, Yalovsky S, Kudla J (2008) Dual fatty acyl modification determines the localization and plasma membrane targeting of CBL/CIPK Ca^{2+} signaling complexes in Arabidopsis. Plant Cell 20:1346–1362
5. Batistic O, Waadt R, Steinhorst L, Held K, Kudla J (2010) CBL-mediated targeting of CIPKs facilitates the decoding of calcium signals emanating from distinct cellular stores. Plant J 61:211–222
6. Chandran V, Stollar EJ, Lindorff-Larsen K, Harper JF, Chazin WJ, Dobson CM, Luisi BF, Christodoulou J (2006) Structure of the regulatory apparatus of a calcium-dependent protein kinase (CDPK): a novel mode of calmodulin-target recognition. J Mol Biol 357:400–410
7. Chen L, Ren F, Zhou L, Wang QQ, Zhong H, Li XB (2012) The Brassica napus calcineurin B-Like 1/CBL-interacting protein kinase 6 (CBL1/CIPK6) component is involved in the plant response to abiotic stress and ABA signalling. J Exp Bot 63:6211–6222
8. Chen X, Gu Z, Xin D, Hao L, Liu C, Huang J, Ma B, Zhang H (2011) Identification and characterization of putative CIPK genes in maize. J Genet Genomics 38:77–87
9. Cuellar T, Azeem F, Andrianteranagna M, Pascaud F, Verdeil JL, Sentenac H, Zimmermann S, Gaillard I (2013) Potassium transport in developing fleshy fruits: the grapevine inward K^+ channel VvK1.2 is activated by CIPK–CBL complexes and induced in ripening berry flesh cells. Plant J 73:1006–1018
10. Cuellar T, Pascaud F, Verdeil JL, Torregrosa L, Adam-Blondon AF, Thibaud JB, Sentenac H, Gaillard I (2010) A grapevine Shaker inward K^+ channel activated by the calcineurin B-like calcium sensor 1-protein kinase CIPK23 network is expressed in grape berries under drought stress conditions. Plant J 61:58–69
11. Deng X, Hu W, Wei S, Zhou S, Zhang F, Han J, Chen L, Li Y, Feng J, Fang B, Luo Q, Li S, Liu Y, Yang G, He G (2013) TaCIPK29, a CBL-interacting protein kinase gene from wheat, confers salt stress tolerance in transgenic tobacco. PLoS ONE 8:e69881
12. Drerup MM, Schlucking K, Hashimoto K, Manishankar P, Steinhorst L, Kuchitsu K, Kudla J (2013) The Calcineurin B-like calcium sensors CBL1 and CBL9 together with their

interacting protein kinase CIPK26 regulate the Arabidopsis NADPH oxidase RBOHF. Mol Plant 6:559–569
13. Du W, Lin H, Chen S, Wu Y, Zhang J, Fuglsang AT, Palmgren MG, Wu W, Guo Y (2011) Phosphorylation of SOS3-like calcium-binding proteins by their interacting SOS2-like protein kinases is a common regulatory mechanism in Arabidopsis. Plant Physiol 156:2235–2243
14. Gao P, Zhao PM, Wang J, Wang HY, Du XM, Wang GL, Xia GX (2008) Co-expression and preferential interaction between two calcineurin B-like proteins and a CBL-interacting protein kinase from cotton. Plant Physiol Biochem 46:935–940
15. Gong D, Guo Y, Jagendorf AT, Zhu JK (2002) Biochemical characterization of the Arabidopsis protein kinase SOS2 that functions in salt tolerance. Plant Physiol 130:256–264
16. Gong D, Guo Y, Schumaker KS, Zhu JK (2004) The SOS3 family of calcium sensors and SOS2 family of protein kinases in Arabidopsis. Plant Physiol 134:919–926
17. Guo L, Yu Y, Xia X, Yin W (2010) Identification and functional characterisation of the promoter of the calcium sensor gene CBL1 from the xerophyte Ammopiptanthus mongolicus. BMC Plant Biol 10:18. doi:10.1186/1471-2229-10-18
18. Halfter U, Ishitani M, Zhu JK (2000) The Arabidopsis SOS2 protein kinase physically interacts with and is activated by the calcium-binding protein SOS3. Proc Natl Acad Sci USA 97:3735–3740
19. Hamada S, Seiki Y, Watanabe K, Ozeki T, Matsui H, Ito H (2009) Expression and interaction of the CBLs and CIPKs from immature seeds of kidney bean (*Phaseolus vulgaris* L.). Phytochemistry 70:501–507
20. Hashimoto K, Eckert C, Anschutz U, Scholz M, Held K, Waadt R, Reyer A, Hippler M, Becker D, Kudla J (2012) Phosphorylation of calcineurin B-like (CBL) calcium sensor proteins by their CBL-interacting protein kinases (CIPKs) is required for full activity of CBL–CIPK complexes toward their target proteins. J Biol Chem 287:7956–7968
21. Hrabak EM, Chan CW, Gribskov M, Harper JF, Choi JH, Halford N, Kudla J, Luan S, Nimmo HG, Sussman MR, Thomas M, Walker-Simmons K, Zhu JK, Harmon AC (2003) The Arabidopsis CDPK-SnRK superfamily of protein kinases. Plant Physiol 132:666–680
22. Hu DG, Li M, Luo H, Dong QL, Yao YX, You CX, Hao YJ (2012) Molecular cloning and functional characterization of MdSOS2 reveals its involvement in salt tolerance in apple callus and Arabidopsis. Plant Cell Rep 31:713–722
23. Hu X, Jiang M, Zhang J, Zhang A, Lin F, Tan M (2007) Calcium-calmodulin is required for abscisic acid-induced antioxidant defense and functions both upstream and downstream of H2O2 production in leaves of maize (Zea mays) plants. New Phytol 173:27–38
24. Imamura M, Yuasa T, Takahashi T, Nakamura N, Hnmp S, Shao-Hui Z, Ken-Ichiro S, Mari II (2008) Isolation and characterization of a cDNA coding cowpea (*Vigna unguiculata* (L.) Walp.) calcineurin B-like protein-interacting protein kinase, VuCIPK1. Plant Biotechnol 25:437–445
25. Ishitani M, Liu J, Halfter U, Kim CS, Shi W, Zhu JK (2000) SOS3 function in plant salt tolerance requires N-myristoylation and calcium binding. Plant Cell 12:1667–1678
26. Chen JH, Sun Y, Sun F, Xia XL, Yin WL (2011) Tobacco plants ectopically expressing the Ammopiptanthus mongolicus AmCBL1 gene display enhanced tolerance to multiple abiotic stresses. Plant Growth Regul 63(3):259–269
27. Kim KN, Cheong YH, Gupta R, Luan S (2000) Interaction specificity of Arabidopsis calcineurin B-like calcium sensors and their target kinases. Plant Physiol 124:1844–1853
28. Kim KN, Lee JS, Han H, Choi SA, Go SJ, Yoon IS (2003) Isolation and characterization of a novel rice Ca^{2+}-regulated protein kinase gene involved in responses to diverse signals including cold, light, cytokinins, sugars and salts. Plant Mol Biol 52:1191–1202
29. Kimura S, Kawarazaki T, Nibori H, Michikawa M, Imai A, Kaya H, Kuchitsu K (2013) The CBL-interacting protein kinase CIPK26 is a novel interactor of Arabidopsis NADPH oxidase AtRbohF that negatively modulates its ROS-producing activity in a heterologous expression system. J Biochem 153:191–195

30. Klee CB, Crouch TH, Krinks MH (1979) Calcineurin: a calcium- and calmodulin-binding protein of the nervous system. Proc Natl Acad Sci USA 76:6270–6273
31. Klee CB, Draetta GF, Hubbard MJ (1988) Calcineurin. Adv Enzymol Relat Areas Mol Biol 61:149–200
32. Knight H, Knight MR (2000) Imaging spatial and cellular characteristics of low temperature calcium signature after cold acclimation in Arabidopsis. J Exp Bot 51:1679–1686
33. Knight H, Trewavas AJ, Knight MR (1996) Cold calcium signaling in Arabidopsis involves two cellular pools and a change in calcium signature after acclimation. Plant Cell 8:489–503
34. Kolukisaoglu U, Weinl S, Blazevic D, Batistic O, Kudla J (2004) Calcium sensors and their interacting protein kinases: genomics of the Arabidopsis and rice CBL–CIPK signaling networks. Plant Physiol 134:43–58
35. Kudla J, Xu Q, Harter K, Gruissem W, Luan S (1999) Genes for calcineurin B-like proteins in Arabidopsis are differentially regulated by stress signals. Proc Natl Acad Sci USA 96:4718–4723
36. Kurusu T, Hamada J, Nokajima H, Kitagawa Y, Kiyoduka M, Takahashi A, Hanamata S, Ohno R, Hayashi T, Okada K, Koga J, Hirochika H, Yamane H, Kuchitsu K (2010) Regulation of microbe-associated molecular pattern-induced hypersensitive cell death, phytoalexin production, and defense gene expression by calcineurin B-like protein-interacting protein kinases, OsCIPK14/15, in rice cultured cells. Plant Physiol 153:678–692
37. Lee KW, Chen PW, Lu CA, Chen S, Ho TH, Yu SM (2009) Coordinated responses to oxygen and sugar deficiency allow rice seedlings to tolerate flooding. Sci Signal 2:ra61
38. Li R, Zhang J, Wu G, Wang H, Chen Y, Wei J (2012) HbCIPK2, a novel CBL-interacting protein kinase from halophyte Hordeum brevisubulatum, confers salt and osmotic stress tolerance. Plant, Cell Environ 35:1582–1600
39. Lin H, Yang Y, Quan R, Mendoza I, Wu Y, Du W, Zhao S, Schumaker KS, Pardo JM, Guo Y (2009) Phosphorylation of SOS3-LIKE CALCIUM BINDING PROTEIN8 by SOS2 protein kinase stabilizes their protein complex and regulates salt tolerance in Arabidopsis. Plant Cell 21:1607–1619
40. Liu J, Zhu JK (1998) A calcium sensor homolog required for plant salt tolerance. Science 280:1943–1945
41. Luan S (2009) The CBL–CIPK network in plant calcium signaling. Trends Plant Sci 14:37–42
42. Luan S, Kudla J, Rodriguez-Concepcion M, Yalovsky S, Gruissem W (2002) Calmodulins and calcineurin B-like proteins: calcium sensors for specific signal response coupling in plants. Plant Cell 14(Suppl):S389–400
43. Luan S, Li W, Rusnak F, Assmann SM, Schreiber SL (1993) Immunosuppressants implicate protein phosphatase regulation of K$^+$ channels in guard cells. Proc Natl Acad Sci USA 90:2202–2206
44. Lyzenga WJ, Liu H, Schofield A, Muise-Hennessey A, Stone SL (2013) Arabidopsis CIPK26 interacts with KEG, components of the ABA signalling network and is degraded by the ubiquitin-proteasome system. J Exp Bot 64:2779–2791
45. Mahajan S, Sopory SK, Tuteja N (2006) Cloning and characterization of CBL–CIPK signalling components from a legume (Pisum sativum). FEBS J 273:907–925
46. Pagnussat GC, Fiol DF, Salerno GL (2002) A CDPK type protein kinase is involved in rice SPS light modulation. Physiol Plant 115:183–189
47. Pandey GK (2008) Emergence of a novel calcium signaling pathway in plants: CBL–CIPK signaling network. Physiol Mol Biol Plants 14:51–68
48. Pardo JM, Reddy MP, Yang S, Maggio A, Huh GH, Matsumoto T, Coca MA, Paino-D'Urzo M, Koiwa H, Yun DJ, Hasegawa PM (1998) Stress signaling through Ca^{2+}-calmodulin-dependent protein phosphatase calcineurin mediates salt adaptation in plants. Proc Natl Acad Sci USA 95:9681–9686
49. Piao HL, Xuan YH, Park SH, Je BI, Park SJ, Kim CM, Huang J, Wang GK, Kim MJ, Kang SM, Lee IJ, Kwon TR, Kim YH, Yeo US, Yi G, Son D, Han CD (2010) OsCIPK31, a

CBL-interacting protein kinase is involved in germination and seedling growth under abiotic stress conditions in rice plants. Mol Cells 30:19–27

50. Pitzschke A, Forzani C, Hirt H (2006) Reactive oxygen species signaling in plants. Antioxid Redox Signal 8:1757–1764
51. Reddy AS (2001) Calcium: silver bullet in signaling. Plant Sci 160:381–404
52. Romeis T, Ludwig AA, Martin R, Jones JD (2001) Calcium-dependent protein kinases play an essential role in a plant defence response. EMBO J 20:5556–5567
53. Rusnak F, Mertz P (2000) Calcineurin: form and function. Physiol Rev 80:1483–1521
54. Sanders D, Pelloux J, Brownlee C, Harper JF (2002) Calcium at the crossroads of signaling. Plant Cell 14(Suppl):S401–417
55. Shang G, Cang H, Liu Z, Gao W, Bi R (2010) Crystallization and preliminary crystallographic analysis of a calcineurin B-like protein 1 (CBL1) mutant from Ammopiptanthus mongolicus. Acta Crystallogr, Sect F: Struct Biol Cryst Commun 66:1602–1605
56. Shao HB, Song WY, Chu LY (2008) Advances of calcium signals involved in plant antidrought. C R Biol 331:587–596
57. Sheen J (1996) Ca^{2+}-dependent protein kinases and stress signal transduction in plants. Science 274:1900–1902
58. Shi J, Kim KN, Ritz O, Albrecht V, Gupta R, Harter K, Luan S, Kudla J (1999) Novel protein kinases associated with calcineurin B-like calcium sensors in Arabidopsis. Plant Cell 11:2393–2405
59. Stewart AA, Ingebritsen TS, Manalan A, Klee CB, Cohen P (1982) Discovery of a Ca^{2+}- and calmodulin-dependent protein phosphatase: probable identity with calcineurin (CaM-BP$_{80}$). FEBS Lett 137:80–84
60. Tai F, Wang Q, Yuan Z, Yuan Z, Li H, Wang W (2013) Characterization of five CIPK genes expressions in maize under water stress. Acta Physiologiae Plantarum 35:1555–1564
61. Tominaga M, Harada A, Kinoshita T, Shimazaki K (2010) Biochemical characterization of calcineurin B-like-interacting protein kinase in Vicia guard cells. Plant Cell Physiol 51:408–421
62. Tripathi V, Parasuraman B, Laxmi A, Chattopadhyay D (2009) CIPK6, a CBL-interacting protein kinase is required for development and salt tolerance in plants. Plant J 58:778–790
63. Tuteja N, Mahajan S (2007) Further characterization of Calcineurin B-like protein and its interacting partner CBL-interacting protein kinase from *Pisum sativum*. Plant Signal Behav 2:358–361
64. Wang M, Gu D, Liu T, Wang Z, Guo X, Hou W, Bai Y, Chen X, Wang G (2007) Overexpression of a putative maize calcineurin B-like protein in Arabidopsis confers salt tolerance. Plant Mol Biol 65:733–746
65. Wang RK, Li LL, Cao ZH, Zhao Q, Li M, Zhang LY, Hao YJ (2012) Molecular cloning and functional characterization of a novel apple MdCIPK6L gene reveals its involvement in multiple abiotic stress tolerance in transgenic plants. Plant Mol Biol 79:123–135
66. Xiang Y, Huang Y, Xiong L (2007) Characterization of stress-responsive CIPK genes in rice for stress tolerance improvement. Plant Physiol 144:1416–1428
67. Yang W, Kong Z, Omo-Ikerodah E, Xu W, Li Q, Xue Y (2008) Calcineurin B-like interacting protein kinase OsCIPK23 functions in pollination and drought stress responses in rice (*Oryza sativa* L.). J Genet Genomics 35:531–543
68. Yim HK, Lim MN, Lee SE, Lim J, Lee Y, Hwang YS (2012) Hexokinase-mediated sugar signaling controls expression of the calcineurin B-like interacting protein kinase 15 gene and is perturbed by oxidative phosphorylation inhibition. J Plant Physiol 169:1551–1558
69. Yu Y, Xia X, Yin W, Zhang H (2007) Comparative genomic analysis of CIPK gene family in Arabidopsis and Populus. Plant Growth Regul 52:101–110
70. Yuasa T, Ishibashi Y, Iwaya-Inoue M (2012) A flower specific calcineurin B-like molecule (CBL)-interacting protein kinase (CIPK) homolog in tomato cultivar micro-tom (*Solanum lycopersicum* L.). AJPS 3:753–763
71. Zhang H, Lv F, Han X, Xia X, Yin W (2013) The calcium sensor PeCBL1, interacting with PeCIPK24/25 and PeCIPK26, regulates Na^+/K^+ homeostasis in *Populus euphratica*. Plant Cell Rep 32:611–621

72. Zhang H, Yin W, Xia X (2008) Calcineurin B-Like family in Populus: comparative genome analysis and expression pattern under cold, drought and salt stress treatment. Plant Growth Regul 56:129–140

73. Zhang H, Yin W, Xia X (2010) Shaker-like potassium channels in Populus, regulated by the CBL–CIPK signal transduction pathway, increase tolerance to low-K^+ stress. Plant Cell Rep 29:1007–1012

74. Zhang H, Yang B, Liu WZ, Li H, Wang L, Wang B, Deng M, Liang W, Deyholos MK, Jiang YQ (2014) Identification and characterization of CBL and CIPK gene families in canola (*Brassica napus* L.). BMC Plant Biol 14. doi:10.1186/1471-2229-14-8

75. Zhang T, Wang Q, Chen X, Tian C, Wang X, Xing T, Li Y, Wang Y (2005) Cloning and bio-chemical properties of CDPK gene OsCDPK14 from rice. J Plant Physiol 162:1149–1159

76. Zhao J, Sun Z, Zheng J, Guo X, Dong Z, Huai J, Gou M, He J, Jin Y, Wang J, Wang G (2009) Cloning and characterization of a novel CBL-interacting protein kinase from maize. Plant Mol Biol 69:661–674

77. Zielinski RE (1998) Calmodulin and calmodulin-binding proteins in plants. Annu Rev Plant Physiol Plant Mol Biol 49:697–725

Chapter 2
Genomic Organization

Abstract After the completion of several plant genomes, *CBL* and *CIPK* genes have been identified in various plant species beside Arabidopsis and rice in angiosperms. Moreover, *CBL* and *CIPK* genes have also been identified in algae, mosses, pteridophytes, and gymnosperms. Interestingly, CBL- and CIPK-type proteins have also been identified outside plant kingdom in protozoa. But, very little is known about the functional role of the CBL and CIPK in these species.

Keywords Genome · Organization · Evolution · CBLs · CIPKs · Gene · Phylogenetic

2.1 Introduction

As organisms become more complex, more numbers of attributes are added to its function. Specificity and cross talks are the most important part of any signaling network in complex genomes; and therefore, it is an interesting question to decipher how signaling molecules and networks are affected during the course of evolution in organisms. In order to answer this question, it is important to understand the genomic organization of a signaling network. CBL and CIPK genes have been identified in algae, mosses, pteridophytes, gymnosperms, and various species of angiosperms (Fig. 2.1) [1, 4, 5, 8, 11, 12]. CIPK constitutes a large gene family in plant genomes compared to CBL. The number of *CBL* and *CIPK* genes increases in the respective gene families from lower plants to higher plants [8]. These findings also suggest that the evolution of the plant lineage coincided with the evolution of complexity of the CBL and CIPK gene families. Duplication of genes played very significant role in the expansion of this gene family. In this chapter, the genomic organization of both *CBL* and *CIPK* gene(s) will be addressed.

© The Author(s) 2014
G.K. Pandey et al., *Global Comparative Analysis of CBL–CIPK Gene Families in Plants*, SpringerBriefs in Plant Science, DOI 10.1007/978-3-319-09078-8_2

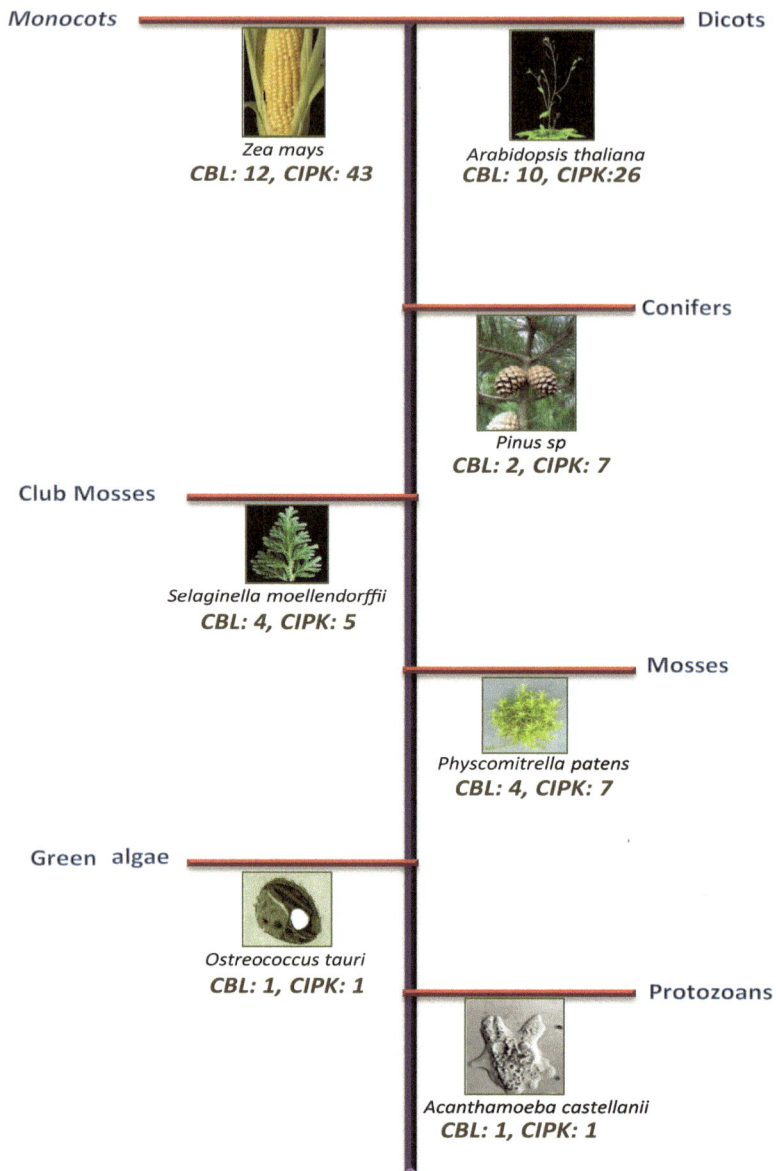

Fig. 2.1 Evolution and complexity of CBLs and CIPKs protein families in different plant lineages. Number of CBL and CIPK genes increases in the respective gene families from lower plants to higher plants suggesting the complexity of the CBL–CIPK system concurrently evolved with the increasing morphological and developmental sophistication of plants

2.2 CBL and CIPK Complements

2.2.1 CBL- and CIPK-Type Proteins in Protozoan

Though, a quest for CBL- and CIPK-related protein in different organism lead to the identification of the calcium signaling module in protozoan eukaryotes such as *Naegleria gruberi* and *Trichomonas vaginalis* [2], but phylogenetic analysis related them more to plant CBL proteins in comparison with animal CNB- and NCS-related proteins [2, 8]. EF-hand 1 was a common structure between protozoans CBL-like and the plants CBLs, similarly, protozoan CIPKs have the characteristic NAF domain-like plant CIPKs. With availability of more genome sequences, CBL and CIPK proteins are also identified in other protozoans like *Acanthamoeba castellanii* and *Emiliania huxleyi* CCMP1516. The presence of both protein families in these non-plant species provides clues for the evolution of CBL–CIPK signaling components.

2.2.2 CBL and CIPK in Algae

A single CBL and CIPK sequences have been identified in single-celled organisms such as *Chlorella* and *Ostreococcus* [2, 8], suggesting early evolution of the CBL–CIPK system at the base of plant lineage. Beside this, CBL–CIPK homologs were discovered from other charophyte green algal species including *Coleochaete orbicularis*, *Lebsormidium flaccidum*, *Chaetospheridium globosum*, and *Penium margaritaceum*. Charophyte *CBL* and *CIPK* homologs displayed expected domain architecture and motifs like their angiosperm counterpart, but these proteins were found to be absent in the algae *Volvox carterii* and *Chlamydomonas rheinhardii*. The exact reason of why CBLs and CIPKs are absent in these algae is not clear; however, these organism might have acquired other Ca^{2+} signaling system for transducing the similar stimuli [2, 8].

2.2.3 CBL and CIPK in Higher Plants

Physcomitrella patens encode four CBL and seven CIPK proteins [8]. The fern *Selaginella moellendorffii* possesses a complement of four *CBLs* and five *CIPKs* [8], whereas conifers like *Pinus* sp. consists of two CBL proteins and seven CIPKs [5]. CBL and CIPK proteins were originally identified in the dicot plant *Arabidopsis thaliana* [6, 7]. After the availability of genome sequence of Arabidopsis, a comprehensive bioinformatic analyses identified a complement of 10 *CBLs* and 26 *CIPKs* [5, 8]. Subsequently, availability of genome sequence in other plants also yield to identify multiple members of CBL–CIPK gene family i.e, 10 genes encoding CBL proteins and 27 genes encoding CIPKs in genome of

Populus trichocarpa (poplar) [8, 11, 12], 8 distinct *CBLs*, and 21 *CIPKs* in *Vitis vinifera* [8]. In monocots, a total of 10 *CBLs* and 30 *CIPKs* were identified in *Oryza sativa* [1, 5]. *Zea mays* has largest CBL and CIPK gene family members with 12 *CBLs* and 43 *CIPKs* in its genome [4]. However, only 6 *CBLs* and 32 CIPK-type kinases were identified in *Sorghum bicolor* [8].

2.3 Genomic Architecture

In the case of multi-gene family, chromosomal distributions of genes on different chromosome provide significant information about the evolution of any gene and gene family in the genome. In the case of genomic distribution of Arabidopsis *CBLs*, most of the genes are present on chromosomes IV and V, whereas in rice, *CBL* genes were located on 6 different chromosomes. As mentioned earlier that the number of *CBLs* and *CIPKs* in the genome increases from lower organism such as protozoa and green algae to higher plants, especially the number of CIPKs genes, so here the role of segmental and tandem duplications might have played a significant role in the expansion of CIPK gene family. In Arabidopsis, there are eight segmental duplication events and two *CIPK* gene pairs arose by tandem duplications [5]. These duplication events are responsible for generation of many homologous gene pairs in the CIPK gene family and amplification of this gene family. Similarly in rice also most of genes arose because of segmental and tandem duplications. According to evolutionary study done by Ye et al. [10] in angiosperm (*O. sativa, Z. mays, A. thaliana,* and *P. trichocarpa*), it was found that gene duplications has played an important role in the expansion of CIPKs.

2.4 Gene Structure

CBLs generally contain multiples introns. In Arabidopsis, rice, *S. bicolor* most of CBL genes harbor seven introns in their coding region. *CIPKs* are divided into two groups based on protein sequence similarity and number of introns, based on later criteria there are two types of CIPK gene groups: intron-rich and intron-poor groups. This kind of grouping has been observed in Arabidopsis, rice, and *Z. mays*. In *Z. mays* out of the total 43 *ZmCIPKs*, 12 *ZmCIPKs* belong to intron-rich group and rest are intron free [4]. In Arabidopsis, out of 25 *CIPK*, eight harbor multiple intron sequences. Rice and poplar also show similar kind of trend. Whereas, in the case of moss *P. patens,* all *CIPKs* contain multiple introns. Furthermore, all the *CIPKs* from the non-angiosperm species were found in intron-rich clade. Analysis of gene duplication showed that the expansion of *CIPKs* (intron-rich and intron-poor) is partly contributed by segmental duplications; however, tandem duplicates were found only in intron-poor *CIPKs* [10].

2.5 Phylogenetic Relatedness and Evolution

CBL number among species varied because of species-specific duplication or deletion genetic modifications [8]. The algal, fern, and moss CBL protein families harbor an N-terminal MGCXXS/T motif that have been shown to allow lipid modification by myristoylation and S-acylation of *A. thaliana* CBL1 [3]. During the evolution, the structural diversification leads to many changes in this membrane-anchored signaling module, which may have resulted in alternative subcellular localizations. Phylogenetic analysis [8] showed that the Arabidopsis CBL10 protein forms a distinct cluster with duplicated CBL protein pairs from poplar, grape, and rice. In general, the comparative analysis of CBLs from all these species further supports the classification of these proteins according to their N-terminal domains.

Ye et al. [10] generated a phylogenetic tree of 146 CIPKs from eight different plant genomes including *Chlorella, Ostreococcus tauri, P. patens, S. moellendorffii, O. sativa, Z. mays, A. thaliana*, and *P. trichocarpa* grouped into two clades, designated as intron-rich clade and intron-less clade. All the CIPKs proteins in algae, moss, and spike moss were grouped into intron-rich clade. According to phylogenetic analysis of various CIPKs of *P. patens, S. moellendorffii, A. thaliana*, and *S. bicolor*, all CIPKs from *P. patens* and *S. moellendorffii* were found to array together with CIPK23 and CIPK24 from *A. thaliana*, shown to represent critical regulators of plant K^+ and Na^+ homeostasis, respectively [9, 13]. This situation may reflect a functional conservation of the CBL–CIPK system in regulating the transport and distribution of these ions in mosses and ferns [8].

References

1. Albrecht V, Ritz O, Linder S, Harter K, Kudla J (2001) The NAF domain defines a novel protein-protein interaction module conserved in Ca^{2+}-regulated kinases. EMBO J 20:1051–1063
2. Batistic O, Kudla J (2009) Plant calcineurin B-like proteins and their interacting protein kinases. Biochim Biophys Acta 1793:985–992
3. Batistic O, Sorek N, Schultke S, Yalovsky S, Kudla J (2008) Dual fatty acyl modification determines the localization and plasma membrane targeting of CBL/CIPK Ca^{2+} signaling complexes in Arabidopsis. Plant Cell 20:1346–1362
4. Chen X, Gu Z, Xin D, Hao L, Liu C, Huang J, Ma B, Zhang H (2011) Identification and characterization of putative CIPK genes in maize. J Genet Genomics 38:77–87
5. Kolukisaoglu U, Weinl S, Blazevic D, Batistic O, Kudla J (2004) Calcium sensors and their interacting protein kinases: genomics of the Arabidopsis and rice CBL–CIPK signaling networks. Plant Physiol 134:43–58
6. Kudla J, Xu Q, Harter K, Gruissem W, Luan S (1999) Genes for calcineurin B-like proteins in Arabidopsis are differentially regulated by stress signals. Proc Natl Acad Sci USA 96:4718–4723
7. Shi J, Kim KN, Ritz O, Albrecht V, Gupta R, Harter K, Luan S, Kudla J (1999) Novel protein kinases associated with calcineurin B-like calcium sensors in Arabidopsis. Plant Cell 11:2393–2405

8. Weinl S, Kudla J (2009) The CBL–CIPK Ca^{2+}-decoding signaling network: function and perspectives. New Phytol 184:517–528

9. Xu J, Li HD, Chen LQ, Wang Y, Liu LL, He L, Wu WH (2006) A protein kinase, interacting with two calcineurin B-like proteins, regulates K^+ transporter AKT1 in Arabidopsis. Cell 125:1347–1360

10. Ye CY, Xia X, Yin W (2013) Evolutionary analysis of CBL-interacting protein kinase gene family in plants. Plant Growth Regul 71:49–56

11. Yu Y, Xia X, Yin W, Zhang H (2007) Comparative genomic analysis of CIPK gene family in Arabidopsis and Populus. Plant Growth Regul 52:101–110

12. Zhang H, Yin W, Xia X (2008) Calcineurin B-Like family in Populus: comparative genome analysis and expression pattern under cold, drought and salt stress treatment. Plant Growth Regulation 56:129–140

13. Zhu JK (2003) Regulation of ion homeostasis under salt stress. Curr Opin Plant Biol 6:441–445

Chapter 3
Distribution and Expression in Plants

Abstract Expression profile of a gene provides clues related to its possible function under a particular condition or stage of growth and development in plants. In many cases, the differential expression of a gene under a particular condition is directly correlated with the function it perform under that condition, however, this is not applicable for many of the genes. Transcriptional profiling and proteomics studies revealed differential expression of several CBLs and CIPKs members under several physiological and developmental stages. The expression profiling of CBLs and CIPKs genes under different conditions and stages of development will enhance our understanding in generating a holistic approach to perform the *in-planta* functional analysis.

Keywords Abiotic stress · Distribution · Expression · Development · Stress markers · Mutant · Overexpression · CBLs · CIPKs

3.1 Introduction

Expression of a gene in a particular plant tissue or stimulus may represent its function during that stage or condition. Majorly, two approaches were employed to determine the expression of *CBLs* and *CIPKs* in plants i.e., RT-PCR and promoter-GUS-based expression analysis. Expression of most of *CBLs* and *CIPKs* is highly affected by abiotic stresses. All *CBLs* and *CIPKs* whose expression have been studied in detail exhibit rather restricted developmental and tissue-specific expression patterns. Same CBL–CIPK complexes can play different functions in specific organs, and specific functions of different CBL–CIPK complexes within a given organ in transducing the calcium-mediated signaling in plants.

G.K. Pandey et al., *Global Comparative Analysis of CBL–CIPK Gene Families in Plants*, SpringerBriefs in Plant Science, DOI 10.1007/978-3-319-09078-8_3

3.2 Distribution of CBLs and CIPKs in Plants

In Arabidopsis, *CBL1* and *CBL9* are expressed in leaf, root, and vascular stele [8, 24, 34]. Promoter-GUS activity of *CBL1* in mature flowers was mainly detected in the anthers especially in the pollen grains [10, 29]. Similarly, a high level of *CBL9* expression was detected in flowers and germinating seeds with *CBL9* promoter::GUS analysis [34]. *CBL3* is expressed at similar levels in leaves, roots, and flowers, whereas *CBL2* is preferentially expressed in roots [24]. GUS activity driven by *CBL2* promoter was seen in the cotyledonary guard cells, in the meristem, and in the elongation zone of roots, the mesophyll cells of rosette leaves, anthers, and stamen filaments of flowers and siliques. Expression of *CBL3* promoter was seen in the root tip and root hair zone, leaf veins, vascular bundles, and the vasculature of sepals, stigma, and receptacle of developing siliques [41]. *SOS3/CBL4* is mainly expressed in root tissues [38, 39]. In contrary, CBL5 was not expressed in the root tissue but significantly expressed in the aerial tissues of the plant [9]. Similarly, in promoter::GUS and real time PCR (RT-PCR) analysis, *SCaBP8/CBL10* reported in low abundance in roots and high level of expression detected in stems [22, 38]. The *AmCBL1* (*Ammopiptanthus mongolicus*) promoter-driven GUS reporter showed expression in tobacco (*Nicotiana tabacum*) leaf veins, stems, and roots, where GUS activity was strictly localized to phloem, and detected in vascular bundles of roots and root tip meristematic zone [17]. Tomato *CBL10* expressed in all the tissues including roots but very low level of expression was detected in stem [11]. Semi-quantitative RT-PCR analysis showed that *Phaseolus vulgaris PvCBL1* is constitutively expressed during seed development and *PvCBL2* also expresses in leaves and early maturing seeds [18]. It is evident from expression analysis that *CBLs* in different plant species are expressed in vascular and meristematic tissue.

In the case of CIPKs, the expression of *CIPK1* was detected in the leaves, and in roots and root tip, and vascular stele [10]. *CIPK3* was found to be expressing uniformly throughout the seedling but a higher level of its expression was detected in the tip of germinating seed radicle [23]. High level of expression detected in stems for *CIPK6* [44]. A GUS activity driven by *CIPK6* promoter detected in cotyledons, leaves, and junction of hypocotyl and roots of 7-day-old seedlings, in stems, leaves, rosettes, flowers, and siliques in mature plant tissues [6].

CIPK8 expression was mainly detected in the roots [14]. In the case of *CIPK9*, a ubiquitous expression was detected in nearly all tissues [33]. Locally restricted expression in leaves has been reported for *CIPK20* [15]. A variable pattern of expression was monitored for *CIPK21* promoter::GUS transgenic lines in Arabidopsis where a high GUS staining was reported for the stem and vascular bundles compared to other part of the plants at the seedling stage (Pandey et al. unpublished data). *SOS2/CIPK24* expressed in both roots and shoots [26, 39], whereas *CIPK14* showed vascular-specific expression [25].

In *Vicia faba*, *VfCIPK1* gene expressed in guard cells and roots [42]. Semi-quantitative RT-PCR analysis showed that the isolated genes *Phaseolus vulgaris*

PVCIPK1-5 expressed in leaves and early maturing seeds [18]. In tomato, *CIPK6* was moderately expressed in all the tissues analyzed except in stems. Histochemical assay revealed that GUS expression driven by *Brassica napus BnCIPK6* promoter in Arabidopsis was high in hypocotyl and cotyledons and no or less expression in rest of the tissues was detected [3].

The overlapping expression patterns of respective CBL–CIPK genes point toward their possible involvement in executing a function at a particular tissue/organ, etc. In an interesting case, CBL1, CBL9, and CIPK23, the expression as reported by the promoter::GUS activity measurement in the different tissues was overlapping in nature [7, 8, 34]. Three of these genes were actively expressed in the stomatal guard cells, in roots, and vascular tissues of leaves [7, 8, 34]. In another example, SOS2/*CIPK24* promoter activity was seen in leaf, stem, flower (similar to *SCABP8/CBL10*), and root tips (similar to *SOS3/CBL4*) [38]. The overlapping expression profiles of *SOS2/CIPK24* with *SOS3/CBL4* and *SCaBP8/CBL10* were also consistent with the interaction between the two CBLs and their target kinase SOS2/CIPK24 [21, 38]. This also hints toward similar roles of SOS3/CBL4 and SCaBP8/CBL10 in different tissue types to activate SOS2/CIPK24 in response to salt stress [38]. Similarly, *CBL1* expression pattern coincide with its interacting protein kinase, *CIPK1* [10]. Root tip particularly at growth zone, vascular tissues of leaves (a strong expression in older leaves), the anthers and the stigma, the overlapping expression profile of *CBL1* and *CIPK1*, hinted their functioning in the same type of plant tissues [10].

3.3 Expressions under Various Environmental and Developmental Conditions

Different CBLs and CIPKs have distinct distribution patterns in plants, and they may function in different pathways. Therefore, the knowledge of expression under different stimuli will help in understanding the functional role of CBL and CIPK in plants. Several reports have been published, ascribing expression of CBLs and CIPKs in several different plants species such as Arabidopsis, rice, *Triticum aestivum*, *Pisum sativum*, and *Malus domestica*, during different environmental stresses.

The expression pattern of *CBL* and *CIPK* genes has been studied extensively in Arabidopsis [2, 24, 27, 32]. Most of the *CBL* and *CIPK* gene expression is regulated under various environmental conditions. Wounding, drought, and cold treatment strongly increased expression of Arabidopsis *CBL1* [24]. Similar to *CBL1*, *CBL9* is also highly inducible by multiple abiotic stresses such as cold, drought, osmotic, and salinity [7, 24, 34]. The level of *CBL1* is not influenced by the exogenous application of abscisic acid (ABA) [1], whereas closely related calcium sensor *CBL9* is induced by ABA [34]. Expression level of *CBL2* and *CBL3* remains unaffected in abiotic stress conditions including, high salt, hyperosmotic stress, or ABA treatment [24, 41], whereas *CBL2* expression was induced by light [31]. No induction was seen in abiotic stress conditions such as high salt, drought, or low temperature in *CBL5* [9]. The promoter activity of *AmCBL1* also positively responds to multiple abiotic stresses such as drought, cold, wounding, salt, $CaCl_2$, ABA, GA, and SA [17].

In Arabidopsis, *CIPK6* gene is induced under multiple stress conditions such as salt, dehydration or drought, and ABA treatment [6]. Significantly enhanced *CIPK6* promoter activity was observed in roots of 11-day-old seedlings treated with salt, mannitol, and ABA [6]. Expression of *CIPK9* is induced mainly by K^+ deficiency conditions and ABA treatment, and abiotic stresses such as NaCl, drought, and wounding [33]. Very recently, our group has identified another CIPK member; *CIPK21* whose expression is induced by ABA, PEG, and mannitol treatment and under stresses such as salt, cold, and drought (G.K. Pandey et al. unpublished data). Expression of *CIPK7* was induced by cold treatment [20]. *SOS2/CIPK24* is up-regulated in response to salt stress [26, 38]. Selective induction of *CIPK* gene expression has also been observed in other plant species.

Interestingly in rice, Chen et al. [4, 5] identified five *OsCIPK* genes (*OsCIPK1*, *OsCIPK2*, *OsCIPK10*, *OsCIPK11,* and *OsCIPK12*) transcriptionally up-regulated after bacterial blight infection *i.e.,* biotic stress responsiveness. The semi-quantitative RT-PCR studies done by Gu et al. [16] on putative rice *CBL* genes under NaCl, PEG, cold, and exogenous ABA treatment and found that expression of all the CBLs was observed in one or more stresses except *OsCBL2* and *OsCBL9*. In another study by Xiang's group [46] shown that different rice *CIPK* genes have different responses to various abiotic stresses. Almost 20 of the rice *CIPK* genes were differentially induced by at least one of the stresses including drought, salinity, cold, polyethylene glycol (PEG), and abscisic acid (ABA) treatment, among them *OsCIPK3* was found to be highly induced by cold stress [46]. In maize, a few of the *ZmCIPK* genes were transcriptionally responsive to abiotic stresses [4, 5]. *ZmCIPK24* was up-regulated by salt stress, while *ZmCIPK31* induced by drought stress and *ZmCIPK20* up-regulated by heat stress and cold stress [4, 5]. One of the CIPK of wheat, *TaCIPK14* was found to be up-regulated under multiple conditions such as cold, salt, PEG, ABA, ethylene, and H_2O_2 [12]. Similarly, *TaCIPK29* transcript was also increased after multiple stress conditions such as NaCl, cold, methyl viologen (MV), abscisic acid (ABA), and ethylene treatments [12]. In woody plants such as apple *SOS2* expression was up-regulated in response to salt stress [19]. As reported by Mahajan et al. [28], transcript level of sweet pea *CBLs* and *CIPKs* were stimulated by increase in the calcium and salicylic acid but did not show any change in transcript accumulation during drought and abscisic acid treatment. Quantitative RT-PCR (qRT-PCR) analysis of canola (*B. napus*) seedlings showed that expression of most of *BnCBLs* and *BnCIPKs* was affected under high salinity, cold, ABA, drought, and oxidative stresses [49]. In Another study, the expression of *BnCIPK6* and *BnCBL1* was significantly up-regulated by salt and osmotic stresses, phosphorous starvation, and abscisic acid (ABA) treatment

[3]. GUS activity driven by *B. napus BnCIPK6* promoter was found to be high in case of salinity, osmotic, and ABA treatment. Expression analysis studies suggest CBL–CIPK network might act dynamically in response to various environmental conditions. In case of *N. tabacum* a homolog of *AtCIPK23* was found to be acclimated during oxidative stress [45]. Overlapping expression in different stimuli was observed in the interacting CBL and CIPK partners such as *CBL2* and *CIPK14* induced by light treatment [31]. Induction of *CBL10* and *CIPK24* takes place under salt-stress condition [26, 38]. Similarly, expression was seen in the case of *CBL9* and *CIPK3* in different abiotic stresses such as salt, drought, cold, and ABA treatment [23, 34].

Furthermore, overlapping or specific expression of *CBLs* and *CIPKs* during different developmental stages has been reported, implicating their function during plant development. *CIPK3* expression is up-regulated especially during the early stages of seedling development [23]. In Arabidopsis, *CBL9*, expression is associated with seed germination and other developmental stages [34]. *CBL1* turned out to be specifically up-regulated during pollen tube growth [29], further *CBL1* promoter-driven GUS activity observed in pollen grains, germinating pollen, and pollen tubes [29]. In *Solanum lycopersicum*, *SlCIPK2* was expressed specifically in the floral organ, particularly in stamen, the flower-specific CIPK was tightly associated with the microsomes [48]. In *Gossypium hirsutum*, *GhCBL2* and *GhCBL3* showed preferential expression in the elongating fiber cells [13] and the expression patterns of these two *CBL* genes coincided with that of a putative CBL-interacting protein kinase gene (*GhCIPK1*) [13]. In some cases, the tissue and developmental-specific expression pattern of *CBL* and *CIPKs* appears to be superimposed by stimulus-specific expression regulation. An overlapping expression pattern was reported by [44], where they showed involvement CIPK6 in root development, as well as in the salt-stress response [44].

3.4 Expression of Stress Markers Genes in Mutant and Overexpressor of CBL and CIPKs

With extensive genetic analysis of mutants and over-expressing transgenic lines of CBLs and CIPKs in Arabidopsis, their role is very well elucidated in various abiotic stresses. Usually, the marker gene expression analysis of mutant and overexpression of a particular gene in plants correlates and explain the molecular nature of the respective phenotype under a particular condition. In the analysis and evaluation of stress response, these marker genes are commonly termed as 'Stress markers'. The changes in kinetics of the stress markers enable to understand the function of a gene in a particular stress response. Most of the stress-marker genes showed very low expression under normal growth condition but when plants are challenged with a particular stress, the expression of stress-marker genes increases tremendously.

Some of the well-studied stress markers are *RD29A, RD29B, RD22, KIN1, RAB18, COR47*, and so on. The stress-marker genes are characterized by presence of one or several stress responsive elements in their promoter region, for example *RD29A* and *RD29B* promoter contains *DREs* and *ABRE* elements and exogenous ABA stimulus induces the expression of both *RD29A* and *RD29B* several fold [30, 43, 50, 51]. The functional role of most of the stress markers has not been known yet but they might be functioning in maintenance of stress-response homeostasis. Based on the differential expression pattern of stress-marker genes in mutant and overexpression transgenic versus wild-type plants, the involvement of a particular gene in regulation of the various stimuli and other abiotic stress response pathways can be ascertained [7, 32, 34, 35, 40, 47].

Cellular calcium regulates various processes including gene expression in plants under biotic and abiotic stresses. Disruption of the *CBL* and *CIPK* gene alters the expression profile of stress-associated genes in most of the cases. In Arabidopsis, the null mutation of *CBL1* made the plants less tolerant to salt and drought stresses but showed enhanced cold tolerance [7]. This altered tolerance to different abiotic stresses was clearly correlated as change in the expression pattern of stress-markers genes in *cbl1* mutants, exhibiting enhanced cold induction and reduced drought induction of stress-marker genes [1, 7], whereas the null mutation of *CBL9* led to induction of stress-marker genes *RAB18, RD22, COR47*, and *COR15A* under ABA- and drought-conditions [34].

In Arabidopsis, *CIPK3* knockout mutants showed decreased expression of *RD29A, RD29B*, and *KIN1* under salt and exogenous ABA treatment, whereas no change in the low temperature-inducible transcription factors (*DREB1A*) was reported [7]. Similarly, *cipk1* mutant is hypersensitive to osmotic and ABA stresses and showed up-regulated expression of stress-marker genes such as *RD29A, KIN1, RD22*, and *RAB18* [10], whereas *CBF1* and *CBF2* expression was unaffected under osmotic stress conditions in *cipk1* mutant [10]. Upon ABA treatment, there is induction of *CBF2* and reduced expression of *KIN1* in *cipk1* mutant [10]. In another study, the alteration of stress-marker genes related to ABA treatment and salt stress was observed in *cipk14* mutant [37]. Hyper induction of stress-marker genes was also reported for *oscipk31::Ds* mutant in rice plants [36].

Overexpression of *CBL1* in Arabidopsis plants led to enhanced salt and drought tolerance but reduced tolerance to freezing [7]. The CBL1-overexpressing plants showed higher expression of most of the stress-marker genes tested under normal condition i.e., without subjecting plants to any stress condition [7]. In the CBL1 overexpressing plants, drought-induced gene expression was enhanced; gene induction by cold was inhibited, which showed an inverse correlation with the expression pattern reported in *cbl1* mutant [7]. Similarly, overexpression of CBL5 also alters the gene expression of *RD29A, RD29B*, and *Kin1* [9]. Constitutively activated kinase, BnCIPK6M (T/D) where there is a mutation in Thr182 and was substituted by Asp in BnCIPK6 when overexpressed in Arabidopsis led to higher expression levels of *ABF3, ABF4*, and *RD29A* than those in the wild type under ABA treatment [3].

References

1. Albrecht V, Weinl S, Blazevic D, D'Angelo C, Batistic O, Kolukisaoglu U, Bock R, Schulz B, Harter K, Kudla J (2003) The calcium sensor CBL1 integrates plant responses to abiotic stresses. Plant J 36:457–470
2. Batistic O, Kudla J (2004) Integration and channeling of calcium signaling through the CBL calcium sensor/CIPK protein kinase network. Planta 219:915–924
3. Chen L, Ren F, Zhou L, Wang QQ, Zhong H, Li XB (2012) The *Brassica napus* calcineurin B-Like 1/CBL-interacting protein kinase 6 (CBL1/CIPK6) component is involved in the plant response to abiotic stress and ABA signalling. J Exp Bot 63:6211–6222
4. Chen X, Gu Z, Xin D, Hao L, Liu C, Huang J, Ma B, Zhang H (2011) Identification and characterization of putative CIPK genes in maize. J Genet Genomics 38:77–87
5. Chen X, Gu Z, Liu F, Ma B, Zhang H (2011) Molecular analysis of rice CIPKs involved in both biotic and abiotic stress responses. Rice Sci 18:1–9
6. Chen L, Wang QQ, Zhou L, Ren F, Li DD, Li XB (2013) Arabidopsis CBL-interacting protein kinase (CIPK6) is involved in plant response to salt/osmotic stress and ABA. Mol Biol Rep 40:4759–4767
7. Cheong YH, Kim KN, Pandey GK, Gupta R, Grant JJ, Luan S (2003) CBL1, a calcium sensor that differentially regulates salt, drought, and cold responses in Arabidopsis. Plant Cell 15:1833–1845
8. Cheong YH, Pandey GK, Grant JJ, Batistic O, Li L, Kim BG, Lee SC, Kudla J, Luan S (2007) Two calcineurin B-like calcium sensors, interacting with protein kinase CIPK23, regulate leaf transpiration and root potassium uptake in Arabidopsis. Plant J 52:223–239
9. Cheong YH, Sung SJ, Kim BG, Pandey GK, Cho JS, Kim KN, Luan S (2010) Constitutive overexpression of the calcium sensor CBL5 confers osmotic or drought stress tolerance in Arabidopsis. Mol Cells 29:159–165
10. D'Angelo C, Weinl S, Batistic O, Pandey GK, Cheong YH, Schultke S, Albrecht V, Ehlert B, Schulz B, Harter K, Luan S, Bock R, Kudla J (2006) Alternative complex formation of the Ca-regulated protein kinase CIPK1 controls abscisic acid-dependent and independent stress responses in Arabidopsis. Plant J 48:857–872
11. de la Torre F, Gutierrez-Beltran E, Pareja-Jaime Y, Chakravarthy S, Martin GB, Del Pozo O (2013) The tomato calcium sensor cbl10 and its interacting protein kinase cipk6 define a signaling pathway in plant immunity. Plant Cell 25:2748–2764
12. Deng X, Hu W, Wei S, Zhou S, Zhang F, Han J, Chen L, Li Y, Feng J, Fang B, Luo Q, Li S, Liu Y, Yang G, He G (2013) TaCIPK29, a CBL-interacting protein kinase gene from wheat, confers salt stress tolerance in transgenic tobacco. PLoS ONE 8:e69881
13. Gao P, Zhao PM, Wang J, Wang HY, Du XM, Wang GL, Xia GX (2008) Co-expression and preferential interaction between two calcineurin B-like proteins and a CBL-interacting protein kinase from cotton. Plant Physiol Biochem 46:935–940
14. Gong D, Gong Z, Guo Y, Chen X, Zhu JK (2002) Biochemical and functional characterization of PKS11, a novel Arabidopsis protein kinase. J Biol Chem 277:28340–28350
15. Gong D, Zhang C, Chen X, Gong Z, Zhu JK (2002) Constitutive activation and transgenic evaluation of the function of an Arabidopsis PKS protein kinase. J Biol Chem 277:42088–42096
16. Gu Z, Ma B, Jiang Y, Chen Z, Su X, Zhang H (2008) Expression analysis of the calcineurin B-like gene family in rice (*Oryza sativa* L.) under environmental stresses. Gene 415:1–12
17. Guo L, Yu Y, Xia X, Yin W (2010) Identification and functional characterisation of the promoter of the calcium sensor gene CBL1 from the xerophyte *Ammopiptanthus mongolicus*. BMC Plant Biol 10:18. doi: 10.1186/1471-2229-10-18
18. Hamada S, Seiki Y, Watanabe K, Ozeki T, Matsui H, Ito H (2009) Expression and interaction of the CBLs and CIPKs from immature seeds of kidney bean (*Phaseolus vulgaris* L.). Phytochemistry 70:501–507
19. Hu DG, Li M, Luo H, Dong QL, Yao YX, You CX, Hao YJ (2012) Molecular cloning and functional characterization of MdSOS2 reveals its involvement in salt tolerance in apple callus and Arabidopsis. Plant Cell Rep 31:713–722

20. Huang C, Ding S, Zhang H, Du H, An L (2011) CIPK7 is involved in cold response by inter-
    acting with CBL1 in *Arabidopsis thaliana*. Plant Sci 181:57–64
21. Ishitani M, Liu J, Halfter U, Kim CS, Shi W, Zhu JK (2000) SOS3 function in plant salt tol-
    erance requires N-myristoylation and calcium binding. Plant Cell 12:1667–1678
22. Kim BG, Waadt R, Cheong YH, Pandey GK, Dominguez-Solis JR, Schultke S, Lee SC,
    Kudla J, Luan S (2007) The calcium sensor CBL10 mediates salt tolerance by regulating ion
    homeostasis in Arabidopsis. Plant J 52:473–484
23. Kim KN, Cheong YH, Grant JJ, Pandey GK, Luan S (2003) CIPK3, a calcium sensor-asso-
    ciated protein kinase that regulates abscisic acid and cold signal transduction in Arabidopsis.
    Plant Cell 15:411–423
24. Kudla J, Xu Q, Harter K, Gruissem W, Luan S (1999) Genes for calcineurin B-like pro-
    teins in Arabidopsis are differentially regulated by stress signals. Proc Natl Acad Sci USA
    96:4718–4723
25. Lee EJ, Iai H, Sano H, Koizumi N (2005) Sugar responsible and tissue specific expres-
    sion of a gene encoding AtCIPK14, an Arabidopsis CBL-interacting protein kinase. Biosci
    Biotechnol Biochem 69:242–245
26. Liu J, Ishitani M, Halfter U, Kim CS, Zhu JK (2000) The Arabidopsis thaliana SOS2
    gene encodes a protein kinase that is required for salt tolerance. Proc Natl Acad Sci USA
    97:3730–3734
27. Luan S (2009) The CBL–CIPK network in plant calcium signaling. Trends Plant Sci
    14:37–42
28. Mahajan S, Sopory SK, Tuteja N (2006) Cloning and characterization of CBL–CIPK signal-
    ling components from a legume (*Pisum sativum*). FEBS J 273:907–925
29. Mahs A, Steinhorst L, Han JP, Shen LK, Wang Y, Kudla J (2013) The calcineurin B-like
    Ca^{2+} sensors CBL1 and CBL9 function in pollen germination and pollen tube growth in
    Arabidopsis. Mol Plant 6:1149–1162
30. Nemhauser JL, Hong F, Chory J (2006) Different plant hormones regulate similar processes
    through largely non-overlapping transcriptional responses. Cell 126:467–475
31. Nozawa A, Koizumi N, Sano H (2001) An Arabidopsis SNF1-related protein kinase, AtSR1,
    interacts with a calcium-binding protein, AtCBL2, of which transcripts respond to light. Plant
    Cell Physiol 42:976–981
32. Pandey GK (2008) Emergence of a novel calcium signaling pathway in plants: CBL–CIPK
    signaling network. Physiol Mol Biol Plants 14:51–68
33. Pandey GK, Cheong YH, Kim BG, Grant JJ, Li L, Luan S (2007) CIPK9: a calcium sen-
    sor-interacting protein kinase required for low-potassium tolerance in Arabidopsis. Cell Res
    17:411–421
34. Pandey GK, Cheong YH, Kim KN, Grant JJ, Li L, Hung W, D'Angelo C, Weinl S, Kudla J,
    Luan S (2004) The calcium sensor calcineurin B-like 9 modulates abscisic acid sensitivity
    and biosynthesis in Arabidopsis. Plant Cell 16:1912–1924
35. Pandey GK, Grant JJ, Cheong YH, Kim BG, Li L, Luan S (2005) ABR1, an APETALA2-
    domain transcription factor that functions as a repressor of ABA response in Arabidopsis.
    Plant Physiol 139:1185–1193
36. Piao HL, Xuan YH, Park SH, Je BI, Park SJ, Kim CM, Huang J, Wang GK, Kim MJ, Kang
    SM, Lee IJ, Kwon TR, Kim YH, Yeo US, Yi G, Son D, Han CD (2010) OsCIPK31, a CBL-
    interacting protein kinase is involved in germination and seedling growth under abiotic stress
    conditions in rice plants. Mol Cells 30:19–27
37. Qin Y, Li X, Guo M, Deng K, Lin J, Tang D, Guo X, Liu X (2008) Regulation of salt and
    ABA responses by CIPK14, a calcium sensor interacting protein kinase in Arabidopsis. Sci
    China C Life Sci 51:391–401
38. Quan R, Lin H, Mendoza I, Zhang Y, Cao W, Yang Y, Shang M, Chen S, Pardo JM, Guo Y
    (2007) SCABP8/CBL10, a putative calcium sensor, interacts with the protein kinase SOS2 to
    protect Arabidopsis shoots from salt stress. Plant Cell 19:1415–1431

39. Shi H, Quintero FJ, Pardo JM, Zhu JK (2002) The putative plasma membrane Na^+/H^+ antiporter SOS1 controls long-distance Na^+ transport in plants. Plant Cell 14:465–477
40. Shinozaki K, Yamaguchi-Shinozaki K (2000) Molecular responses to dehydration and low temperature: differences and cross-talk between two stress signaling pathways. Curr Opin Plant Biol 3:217–223
41. Tang RJ, Liu H, Yang Y, Yang L, Gao XS, Garcia VJ, Luan S, Zhang HX (2012) Tonoplast calcium sensors CBL2 and CBL3 control plant growth and ion homeostasis through regulating V-ATPase activity in Arabidopsis. Cell Res 22:1650–1665
42. Tominaga M, Harada A, Kinoshita T, Shimazaki K (2010) Biochemical characterization of calcineurin B-like-interacting protein kinase in vicia guard cells. Plant Cell Physiol 51:408–421
43. Toufighi K, Brady SM, Austin R, Ly E, Provart NJ (2005) The botany array resource: e-northerns, expression angling, and promoter analyses. Plant J. 43:153–163
44. Tripathi V, Parasuraman B, Laxmi A, Chattopadhyay D (2009) CIPK6, a CBL-interacting protein kinase is required for development and salt tolerance in plants. Plant J 58:778–790
45. Vranová E, Atichartpongkul S, Villarroel R, Van Montagu M, Inzé D, Van Camp W (2002) Comprehensive analysis of gene expression in *Nicotiana tabacum* leaves acclimated to oxidative stress. Proc Natl Acad Sci USA 99:10870–10875
46. Xiang Y, Huang Y, Xiong L (2007) Characterization of stress-responsive CIPK genes in rice for stress tolerance improvement. Plant Physiol 144:1416–1428
47. Yamaguchi-Shinozaki K, Shinozaki K (1994) A novel cis-acting element in an Arabidopsis gene is involved in responsiveness to drought, low-temperature, or high-salt stress. Plant Cell 6:251–264
48. Yuasa T, Ishibashi Y, Iwaya-Inoue M (2012) A flower specific calcineurin B-like molecule (CBL)-interacting protein kinase (CIPK) homolog in tomato cultivar micro-tom (*Solanum lycopersicum* L.). AJPS 3:753–763
49. Zhang H, Yang B, Liu W-Z, Li H, Wang L, Wang B, Deng M, Liang W, Deyholos MK, Jiang Y-Q (2014) Identification and characterization of CBL and CIPK gene families in canola (*Brassica napus* L.). BMC Plant Biol 14:8. doi:10.1186/1471-2229-14-8
50. Zimmermann P, Hennig L, Gruissem W (2005) Gene-expression analysis and network discovery using Genevestigator. Trends Plant Sci 10:407–409
51. Zimmermann P, Hirsch-Hoffmann M, Hennig L, Gruissem W (2004) Genevestigator. Arabidopsis microarray database and analysis toolbox. Plant Physiol 136:2621–2632

Chapter 4
Protein Structure and Localization

Abstract There are multiple calcium-binding CBLs and their interacting kinases, CBL-interacting protein kinases (CIPKs), present in the plants. Available structural information provides a general regulatory mechanism by which CBLs and CIPKs function together in decoding calcium signals elicited by different environmental stimuli. A cellular localization matrix of a plant CBL and CIPK signaling network with their molecular mechanism of structural regulation will be significantly helpful in understanding the specificity of cellular responses during plant calcium signaling network in different intra- and extracellular stores.

Keywords Protein · Structure · Motifs · CBL–CIPK complexes · Subcellular localization · BiFC · Targeting

4.1 Introduction

One of the important components in calcium signaling pathway is calcium sensor, which bind to calcium ion followed by a conformational change responsible for binding and activating its interacting proteins [10, 18]. Most of the calcium sensors possess EF-hands to bind Ca^{2+} [7, 11], composed of a helix–turn–helix structural domain found in a large family of calcium-binding proteins. Ca^{2+} has the ability to coordinate with six to eight uncharged oxygen atoms in the EF-hand domain [12]. EF-hand consists of two alpha helices linked by a short-loop region that usually binds calcium ions [3, 14, 16]. The EF-hand structural motifs, sequence requirements were analyzed in detail. The calcium ion is coordinated in a pentagonal bipyramidal configuration. The six residues involved in the binding are in positions 1, 3, 5, 7, 9, and 12; these residues are denoted by X, Y, Z, $-Y$, $-X$, and $-Z$. Out of these, five residue bind calcium preferably to oxygen-containing side chains, especially aspartate and glutamate, whereas sixth residue is necessarily be glycine. Binding of Ca^{2+} leads to conformational changes in the CBL, which allow its efficient binding and activation of CIPK protein [29].

© The Author(s) 2014
G.K. Pandey et al., *Global Comparative Analysis of CBL–CIPK Gene Families in Plants*, SpringerBriefs in Plant Science, DOI 10.1007/978-3-319-09078-8_4

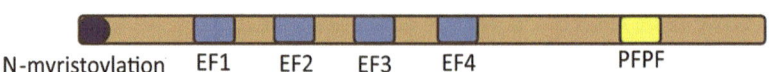

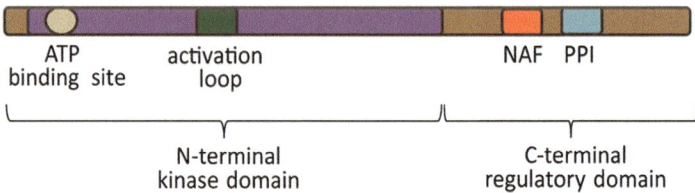

Fig. 4.1 Various motifs and domains in the CBL and CIPK. CBLs contain a conserved N-myristoylation motif shown as *blue round box*. *Light blue rectangle* represents four EF-hand motifs and *yellow rectangle* represents PFPF motif. CIPK proteins contain a conserved N-terminal catalytic kinase domain (*light purple box*), which have ATP binding site (*light yellow round box*) and activation loop (*green rectangle box*). The C-terminal regulatory domain contains FISL or NAF motifs (*orange box*) and PPI motif (*light blue box*)

4.2 Motifs and Domains

4.2.1 Motifs in the CBLs

All CBL proteins share a rather conserved core region consisting of four EF-hand calcium-binding domains, separated by spacing regions encompassing an absolutely conserved number of amino acids in all CBL proteins (Fig. 4.1). No CBL protein contains four canonical EF-hand motifs; therefore, none of the CBL binds to four Ca^{2+} ions. In Arabidopsis, CBL2, CBL3, CBL4, CBL5, and CBL8 possess no canonical EF-hands whereas CBL6, CBL7, and CBL10 possess one; and CBL1 and CBL9 contain two canonical EF-hands [3]. Very interestingly, some of the members of the whole CBL component have N-terminal MGCXXS/T motif. This motif is responsible for lipid modification by myristoylation and S-acylation for membrane association [4, 6]. In most of the plant genomes like algae, fern, moss, and higher plants like Arabidopsis and rice, many members of CBL protein family harbor an N-terminal MGCXXS/T motif responsible for lipid modification by myristoylation and S-acylation [4–6, 41].

In Arabidopsis, four CBLs possess this motif and, in rice, five of the CBLs contained this motif. CBLs also possess motif called 'PFPF motif' at the C-terminus. CBL-interacting protein kinase (CIPKs) phosphorylates CBL at the conserved Ser residue of the PFPF motif (Fig. 4.1) [11, 19]. All CBL proteins are more or less similar in size, i.e., usually 23–26 kDa. CBL proteins can be divided into two groups according to the N-terminal domains: CBL proteins with short N-terminal domain of 27–32 amino acids and CBL proteins with an extended N-terminus (41–43 amino acids). The short N-terminus-containing group comprises CBL1, CBL4, CBL5, CBL8, and CBL9 in Arabidopsis [4] and OsCBL1, OsCBL4, OsCBL5, OsCBL7, and OsCBL8 in rice. Whereas CBL2, CBL3, and CBL6 in Arabidopsis, and OsCBL2 and OsCBL3 in rice form the group of CBL proteins with an extended N-terminus. With the exception of CBL10 in Arabidopsis [24] and OsCBL6, OsCBL9, and OsCBL10 in case of rice [23], which harbors an unusually large N-terminus.

4.2.2 Motifs in the CIPKs

CIPKs belong to sucrose non-fermenting 1 (SNF1)-related kinases, group 3 (SnRK3) [21]. CIPK protein is divided into two domains, a catalytic domain and C-terminal regulatory domain. CIPK's catalytic domain has serine/threonine kinase motif. The regulatory domain has a conserved 21-amino acid FISL (also known as NAF) motif and the PPI domain (Fig. 4.1). The binding of CIPK to CBL is mediated by the FISL motif [2], which is auto-inhibitory to the kinase activity whereas PPI binding motif is responsible for interaction with PP2C [28]. Additionally, phosphorylation within the activation loop of the kinase domain by yet unknown kinase enhances the enzymatic activity of CIPKs and leads to CBL-independent auto-phosphorylation of these proteins [3, 13, 15, 29]. Similar to all other kinases, CIPKs also have a conserved ATP binding site in the catalytic kinase domain.

4.3 Protein Structure

4.3.1 Three-Dimensional Structure of CBLs

Nagae et al. [27], first reported the crystal structure of CBL2, a CBL family member. From the crystal structure of CBL2, it was revealed that CBL2 is comprised of two globular structures, separated by a short linker domain that shares overall topology with calcineurin-B (CNB) and neuronal calcium sensor (NCS). α-helical structure composed of nine α-helices (αA–αI), two 3_{10}-helices (αJ and αK) and four short β-strands [27]. CBLs have four EF-hand motifs comprised of helix–loop–helix structure responsible for calcium binding. A canonical EF-hand is characterized by a sequence with the pattern that participates in calcium coordination. The two calcium ions are coordinated in the first and fourth EF-hand motifs, whereas the second and third EF-hand motifs are maintained in the open form by internal hydrogen bonding without coordination of calcium ions. Such differences in calcium coordination within the CBL family might be related to the specificity of target recognition in calcium signaling.

4.3.2 CBL–CIPK Complexes

No structural information is available about the catalytic domain of CIPKs; the Ser/Thr protein kinase domain within the catalytic domain may be similar to the SNF1 kinase domains [33, 40]. The crystal structures of CBL2 and SOS3/CBL4 with regulatory domain of CIPK14 and SOS2/CIPK24, respectively, are very well studied, which helped in understanding many regulatory questions of CBL–CIPK networking [1, 32]. It can be implied from the two studies that CBL provides a hydrophobic crevice where the auto-inhibitory NAF domain of CIPK binds. The rest of the regulatory C-terminal of the kinase along with the PPI domain is removed from the catalytic domain and the kinase is activated. The PPI domain is so placed that it makes kinase activation and phosphatase binding to PPI domain a mutually exclusive event [33]. Large numbers of hydrophobic interactions are responsible for recognition and maintenance of CBLs and the helical FISL motif complex and mainly responsible for the stabilization of the complex [33]. This kind of the structural analysis help in solving many regulatory aspects of CBL–CIPK networking, such as the molecular basis of the selectivity of CBL–CIPK interaction. The specificity of CBL–CIPK interaction is also determined by the divergent nature of FISL motif, which showed different level of divergence at the loop connecting the two helical segments of FISL only [1, 33].

4.4 Subcellular Localization

To understand the function of CBL–CIPK complex and its specificity in the Ca^{2+} signaling system, it is very important to understand the subcellular localization of CBL and CIPK spatially and temporally. By differential subcellular localization of CBL–CIPK complexes in the cell such as on the plasma membrane, endomembranes, and other organelles, a distinct function can be achieved in a particular stimulus–response coupling reaction. Moreover, by typical subcellular targeting of CBL–CIPK complexes, specificity in signaling pathways can be achieved. Subcellular localization of many of the CBL–CIPK complexes has been studied in Arabidopsis and rice. Most of the subcellular location of CBLs in the cell is distinct whereas many of the CIPKs are distributed throughout the cell. In Arabidopsis, most of the in vivo localization studies using bimolecular fluorescence complementation (BiFC) have proved that CBLs are responsible for localization of the interacting CIPK at different sites within the cell, where they can interact and modulate various target proteins.

4.4.1 Subcellular Localization of CBL Gene Family

Since many of CBLs family members in plants (from algae to higher plants) possess the N-terminal MGCXXS/T motif (myristoylation and S-acylation conserved motif). This motif is responsible for lipid modification by myristoylation and S-acylation of

CBLs for membrane association [4]. In Arabidopsis, four CBLs, i.e., CBL1, CBL4, CBL5, and CBL9 harbor conserved myristoylation motif and hence are targeted to plasma membrane. CBL proteins decode calcium signals not only at the plasma membrane (CBL1 and CBL9) and the tonoplast (CBL2, CBL6, CBL10, and CBL3), but also in the cytoplasm and nucleus (CBL7 and CBL8). Taken together, it can be hypothesized that CBL1, CBL4, and CBL9 predominantly fulfill their function at the plasma membrane, and CBL4 and CBL5 could in addition also target proteins in the cytoplasm and nucleus [5]. In the case of rice, OsCBL2 and OsCBL3 were localized to the tonoplast and OsCBL4 was localized to the plasma membrane [22]. In vivo targeting experiments revealed that OsCBL8 localized to the plasma membrane [17]. In *Zea mays*, ZmCBL3, ZmCBL4, and ZmCBL5 were localized to the plasma membrane [42]. Moreover, N-myristoylation motif was identified in ZmCBL3 and ZmCBL4, and in the case of ZmCBL3, several palmitoylation sites were also considered to be responsible for their membrane localization [42].

4.4.2 Subcellular Localization of CIPK Gene Family

In Arabidopsis, most of CIPKs showed a cytoplasmic or nucleoplasmic accumulation. CIPK1, CIPK23, and CIPK24 were localized throughout including the cytoplasmic and nuclear compartments [8, 9, 24]. GFP fusion proteins of CIPK1, CIPK2, CIPK3, CIPK4, CIPK7, CIPK8, CIPK10, CIPK14, CIPK17, CIPK23, and CIPK24 displayed a similar localization pattern that included strong fluorescence in the cytoplasm and nucleus [5]. In Zea *mays*, ZmCIPK16 was mainly localized at the nucleus and the plasma membrane, and also at a low level in the cytosol [42]. Subcellular localization of CIPK from *Triticum aestivum*, TaCIPK14, was found to be throughout the cell. In *Vicia faba*, VfCIPK1 protein localized on the outer membrane of mitochondria in guard cells [35].

4.4.3 Subcellular Targeting of CBL–CIPK Complexes

As mentioned above, most of the location of CBLs in the cell is distinct whereas many of the CIPKs are distributed throughout the cell. Therefore, an important question arises, how CBL–CIPK complexes are localized in the cell to perform specific and broader functions.

In most cases, cell perceives the external stimuli at the plasma membrane and leads to generation of second messenger, for example, transient increase in Ca^{2+} concentration also designated as 'calcium-signature,' which is further decoded by calcium sensors such as CBLs. Calcium-bound CBL is considered active in nature and hence further activates its binding partners, i.e., CIPKs and catalyzes the phosphorylation reaction of specific target proteins in the signaling pathway. Despite no recognizable localization, motif for CIPKs has been found till date [25].

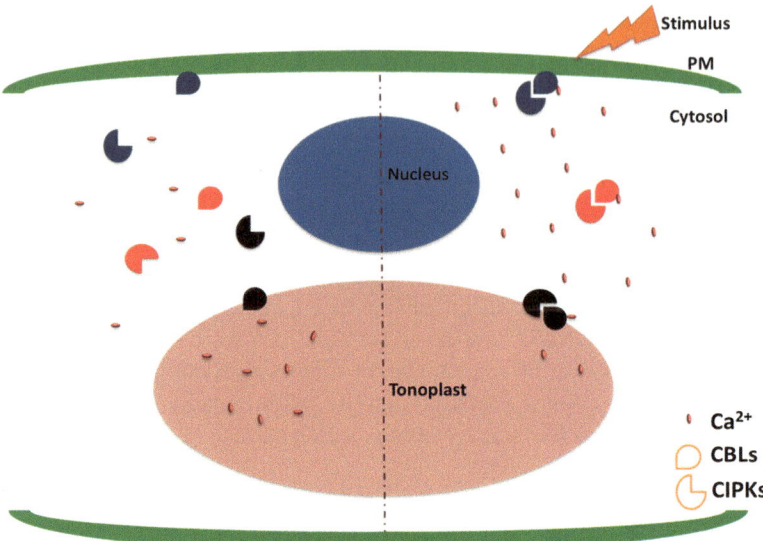

Fig. 4.2 Subcellular targeting of CBL–CIPK complexes. A stimulus is perceived; calcium release from various cellular stores result in transient increase in the cytoplasmic calcium level, which binds and activates CBLs. Activated CBLs target CIPKs to the various site of action like plasma membrane, tonoplast and cytoplasm

In Arabidopsis, most of the in vivo localization by BiFC have proved that CBLs are responsible for localization of the interacting CIPKs at different sites within the cell (Fig. 4.2) [3, 5]. One of the well-known examples is plasma membrane targeting of CIPK1, CIPK23, and CIPK24 by CBL1 and CBL9 [9, 37, 39]. Similarly, targeting of SOS2/CIPK24 by SOS3/CBL4 in salt overly sensitive (SOS) pathway at the plasma membrane where SOS3/CBL4–SOS2/CIPK24 complex can phosphorylate sodium-proton antiporter SOS1 [31]. A special case was observed for the interaction of the cytosolic calcium sensor CBL8 with the kinase CIPK14. Interaction between these two proteins was exclusively detected at the plasma membrane [5]. Considering that CBL8 is not lipid modified at its N-terminus and thereby differs from the lipid-modified plasma membrane targeted CBL1, 4, 5 and 9; and this points toward yet unidentified targeting mechanism of CBL8–CIPK14 complexes. For some of the CIPKs, i.e., CIPK9, CIPK14, their interacting CBLs (CBL2 and CBL3) are tonoplast localized, which target these CIPKs to the vacuolar membrane *in-planta*, indicating that one and the same kinase can be recruited to different subcellular destinations by interacting with different CBLs located at different membranes [5, 26].

CBL4–CIPK6–AKT2 targeting to plasma membrane where CBL4–CIPK6 complex is required to recruits AKT2 efficiently to plasma membrane without phosphorylating AKT2 [20]. Concurrent interactions of some CIPKs have also been seen with different CBLs on different sites in the cells. CBL8 and CIPK14 complex formed exclusively at the plasma membrane, however, CBL2 and CBL3 target CIPK14 to the vacuolar membrane [34, 36, 38]. This indicates that CIPK14, depending on the identity of its interacting CBL protein, represents a dual-functioning protein kinase that regulates targets in different compartments of the cell. Similar dual function of kinases was also seen in the case of concurrent interaction of the protein kinase CIPK24 with the calcium sensors CBL4 and CBL10 at the plasma membrane and tonoplast, respectively, in regulation of salt stress tolerance [24, 30]. The CBL–CIPK complex formation was also found to be localized spatially in other plant species. In *Zea mays*, ZmCIPK16 recruits ZmCBL3, ZmCBL4, and ZmCBL5 to form complex on the plasma membrane. Similarly, ZmCIPK16 was found to be interacting with ZmCBL3, ZmCBL4, ZmCBL5, and ZmCBL8 at varying strength [42]. ZmCIPK16 was localized to the nucleus, cell membrane, and to a lesser extent in the cytosol. The ZmCBL3, 4, and 5 were found to be restricted to the plasma membrane, and the complex of these maize CBLs with ZmCIPK16 was found to be plasma membrane localized [42]. ZmCBL3–ZmCIPK16, ZmCBL4–ZmCIPK16, and ZmCBL5–ZmCIPK16 complexes might interact and phosphorylate some membrane-localized proteins such as transporters/channels in the signaling cascade. Among these, the strongest interaction was observed between ZmCBL4 and ZmCIPK16, which perhaps indicate the role of both these interacting partners in the salt stress signaling, as reported earlier for ZmCBL4 expression led to improve salt tolerance [42]. Overall, in most of the cases, CBLs determined the preferential or differential targeting of CIPKs spatially to different location and hence regulate the function imparted by CBL–CIPK complex in a typical stimulus–response coupling process.

References

1. Akaboshi M, Hashimoto H, Ishida H, Saijo S, Koizumi N, Sato M, Shimizu T (2008) The crystal structure of plant-specific calcium-binding protein AtCBL2 in complex with the regulatory domain of AtCIPK14. J Mol Biol 377:246–257
2. Albrecht V, Ritz O, Linder S, Harter K, Kudla J (2001) The NAF domain defines a novel protein-protein interaction module conserved in Ca^{2+}-regulated kinases. EMBO J 20:1051–1063
3. Batistic O, Kudla J (2004) Integration and channeling of calcium signaling through the CBL calcium sensor/CIPK protein kinase network. Planta 219:915–924
4. Batistic O, Sorek N, Schultke S, Yalovsky S, Kudla J (2008) Dual fatty acyl modification determines the localization and plasma membrane targeting of CBL/CIPK Ca^{2+} signaling complexes in Arabidopsis. Plant Cell 20:1346–1362
5. Batistic O, Waadt R, Steinhorst L, Held K, Kudla J (2010) CBL-mediated targeting of CIPKs facilitates the decoding of calcium signals emanating from distinct cellular stores. Plant J 61:211–222
6. Batistic O, Rehers M, Akerman A, Schlucking K, Steinhorst L, Yalovsky S, Kudla J (2012) S-acylation-dependent association of the calcium sensor CBL2 with the vacuolar membrane is essential for proper abscisic acid responses. Cell Res 22:1155–1168
7. Burgoyne RD (2007) Neuronal calcium sensor proteins: generating diversity in neuronal Ca^{2+} signalling. Nat Rev Neurosci 8:182–193
8. Cheong YH, Pandey GK, Grant JJ, Batistic O, Li L, Kim BG, Lee SC, Kudla J, Luan S (2007) Two calcineurin B-like calcium sensors, interacting with protein kinase CIPK23, regulate leaf transpiration and root potassium uptake in Arabidopsis. Plant J 52:223–239

9. D'Angelo C, Weinl S, Batistic O, Pandey GK, Cheong YH, Schultke S, Albrecht V, Ehlert B, Schulz B, Harter K, Luan S, Bock R, Kudla J (2006) Alternative complex formation of the Ca-regulated protein kinase CIPK1 controls abscisic acid-dependent and independent stress responses in Arabidopsis. Plant J 48:857–872

10. Das R, Pandey GK (2010) Expressional analysis and role of calcium regulated kinases in abiotic stress signaling. Curr Genomics 11:2–13

11. Du W, Lin H, Chen S, Wu Y, Zhang J, Fuglsang AT, Palmgren MG, Wu W, Guo Y (2011) Phosphorylation of SOS3-like calcium-binding proteins by their interacting SOS2-like protein kinases is a common regulatory mechanism in Arabidopsis. Plant Physiol 156:2235–2243

12. Fedrizzi L, Lim D, Carafoli E (2008) Calcium and signal transduction. Biochem Mol Biol Educ 36:175–180

13. Fujii H, Zhu JK (2009) An autophosphorylation site of the protein kinase SOS2 is important for salt tolerance in Arabidopsis. Mol Plant 2:183–190

14. Gifford JL, Walsh MP, Vogel HJ (2007) Structures and metal-ion-binding properties of the Ca^{2+}-binding helix–loop–helix EF-hand motifs. Biochem J 405:199–221

15. Gong D, Guo Y, Jagendorf AT, Zhu JK (2002) Biochemical characterization of the Arabidopsis protein kinase SOS2 that functions in salt tolerance. Plant Physiol 130:256–264

16. Gong D, Guo Y, Schumaker KS, Zhu JK (2004) The SOS3 family of calcium sensors and SOS2 family of protein kinases in Arabidopsis. Plant Physiol 134:919–926

17. Gu Z, Ma B, Jiang Y, Chen Z, Su X, Zhang H (2008) Expression analysis of the calcineurin B-like gene family in rice (*Oryza sativa* L.) under environmental stresses. Gene 415:1–12

18. Harmon AC, Gribskov M, Gubrium E, Harper JF (2001) The CDPK superfamily of protein kinases. New Phytol 151:175–183

19. Hashimoto K, Eckert C, Anschutz U, Scholz M, Held K, Waadt R, Reyer A, Hippler M, Becker D, Kudla J (2012) Phosphorylation of calcineurin B-like (CBL) calcium sensor proteins by their CBL-interacting protein kinases (CIPKs) is required for full activity of CBL–CIPK complexes toward their target proteins. J Biol Chem 287:7956–7968

20. Held K, Pascaud F, Eckert C, Gajdanowicz P, Hashimoto K, Corratge-Faillie C, Offenborn JN, Lacombe B, Dreyer I, Thibaud JB, Kudla J (2011) Calcium-dependent modulation and plasma membrane targeting of the AKT2 potassium channel by the CBL4/CIPK6 calcium sensor/protein kinase complex. Cell Res 21:1116–1130

21. Hrabak EM, Chan CW, Gribskov M, Harper JF, Choi JH, Halford N, Kudla J, Luan S, Nimmo HG, Sussman MR, Thomas M, Walker-Simmons K, Zhu JK, Harmon AC (2003) The Arabidopsis CDPK-SnRK superfamily of protein kinases. Plant Physiol 132:666–680

22. Hwang YS, Bethke PC, Cheong YH, Chang HS, Zhu T, Jones RL (2005) A gibberellin-regulated calcineurin B in rice localizes to the tonoplast and is implicated in vacuole function. Plant Physiol 138:1347–1358

23. Kanwar P, Sanyal SK, Tokas I, Yadav AK, Pandey A, Kapoor S, Pandey GK (2014) Comprehensive structural, interaction and expression analysis of CBL and CIPK complement during abiotic stresses and development in rice. Cell calcium (in press)

24. Kim BG, Waadt R, Cheong YH, Pandey GK, Dominguez-Solis JR, Schultke S, Lee SC, Kudla J, Luan S (2007) The calcium sensor CBL10 mediates salt tolerance by regulating ion homeostasis in Arabidopsis. Plant J 52:473–484

25. Kolukisaoglu U, Weinl S, Blazevic D, Batistic O, Kudla J (2004) Calcium sensors and their interacting protein kinases: genomics of the Arabidopsis and rice CBL–CIPK signaling networks. Plant Physiol 134:43–58

26. Liu LL, Ren HM, Chen LQ, Wang Y, Wu WH (2013) A protein kinase, calcineurin B-like protein-interacting protein Kinase9, interacts with calcium sensor calcineurin B-like Protein3 and regulates potassium homeostasis under low-potassium stress in Arabidopsis. Plant Physiol 161:266–277

27. Nagae M, Nozawa A, Koizumi N, Sano H, Hashimoto H, Sato M, Shimizu T (2003) The crystal structure of the novel calcium-binding protein AtCBL2 from Arabidopsis thaliana. J Biol Chem 278:42240–42246

28. Ohta M, Guo Y, Halfter U, Zhu JK (2003) A novel domain in the protein kinase SOS2 mediates interaction with the protein phosphatase 2C ABI2. Proc Natl Acad Sci USA 100:11771–11776
29. Pandey GK (2008) Emergence of a novel calcium signaling pathway in plants: CBL–CIPK signaling network. Physiol Mol Biol Plants 14:51–68
30. Quan R, Lin H, Mendoza I, Zhang Y, Cao W, Yang Y, Shang M, Chen S, Pardo JM, Guo Y (2007) SCABP8/CBL10, a putative calcium sensor, interacts with the protein kinase SOS2 to protect Arabidopsis shoots from salt stress. Plant Cell 19:1415–1431
31. Quintero FJ, Ohta M, Shi H, Zhu JK, Pardo JM (2002) Reconstitution in yeast of the Arabidopsis SOS signaling pathway for Na+ homeostasis. Proc Natl Acad Sci USA 99:9061–9066
32. Sanchez-Barrena MJ, Fujii H, Angulo I, Martinez-Ripoll M, Zhu JK, Albert A (2007) The structure of the C-terminal domain of the protein kinase AtSOS2 bound to the calcium sensor AtSOS3. Mol Cell 26:427–435
33. Sanchez-Barrena MJ, Martinez-Ripoll M, Albert A (2013) Structural biology of a major signaling network that regulates plant abiotic stress: The CBL–CIPK mediated pathway. Int J Mol Sci 14:5734–5749
34. Steinhorst L, Kudla J (2013) Calcium and reactive oxygen species rule the waves of signaling. Plant Physiol 163:471–485
35. Tominaga M, Harada A, Kinoshita T, Shimazaki K (2010) Biochemical characterization of calcineurin B-like-interacting protein kinase in Vicia guard cells. Plant Cell Physiol 51:408–421
36. Waadt R, Kudla J (2008) In planta visualization of protein interactions using bimolecular fluorescence complementation (BiFC). CSH Protoc, pdb.prot4995
37. Waadt R, Schmidt LK, Lohse M, Hashimoto K, Bock R, Kudla J (2008) Multicolor bimolecular fluorescence complementation reveals simultaneous formation of alternative CBL/CIPK complexes in planta. Plant J 56:505–516
38. Walter M, Chaban C, Schutze K, Batistic O, Weckermann K, Nake C, Blazevic D, Grefen C, Schumacher K, Oecking C, Harter K, Kudla J (2004) Visualization of protein interactions in living plant cells using bimolecular fluorescence complementation. Plant J 40:428–438
39. Xu J, Li HD, Chen LQ, Wang Y, Liu LL, He L, Wu WH (2006) A protein kinase, interacting with two calcineurin B-like proteins, regulates K+ transporter AKT1 in Arabidopsis. Cell 125:1347–1360
40. Yunta C, Martínez-Ripoll M, Zhu JK, Albert A (2011) The structure of Arabidopsis thaliana OST1 provides insights into the kinase regulation mechanism in response to osmotic stress. J Mol Biol 414:135–144
41. Zhang H, Yang B, Liu W-Z, Li H, Wang L, Wang B, Deng M, Liang W, Deyholos MK, Jiang Y-Q (2014) Identification and characterization of CBL and CIPK gene families in canola (Brassica napus L.). BMC Plant Biol 14. doi:10.1186/1471-2229-14-8
42. Zhao J, Sun Z, Zheng J, Guo X, Dong Z, Huai J, Gou M, He J, Jin Y, Wang J, Wang G (2009) Cloning and characterization of a novel CBL-interacting protein kinase from maize. Plant Mol Biol 69:661–674

Chapter 5
Biochemical Properties of CBLs and CIPKs

Abstract One of the important aspects of calcium sensor CBL is that they can bind and decode calcium signature generated in the signaling pathway. As reflected from the biochemical nature of CBLs, they do not possess any catalytic activity and hence can only act as signal relay in a pathway by binding Ca^{2+}, whereas the CBL-interacting protein kinase (CIPKs) possess the catalytic kinase activity and phosphorylate its targeted substrate. In nutshell, it is concluded that CBL can bind calcium ion and this binding of calcium will lead to change in confirmation in such a way that Ca^{2+}–CBL can further activate its interacting protein, i.e., a protein kinases designated as CIPKs.

Keywords Biochemical · CBLs · CIPKs · Kinase · Target/substrate · Phosphorylation

5.1 Introduction

Since CBL-interacting protein kinase (CIPKs) possess enzymatic kinase activity, most of these kinases (as an enzyme) comprises of several physical and biochemical properties such as divalent cation preference, phosphate donor specificity, steady-state substrate kinetics, and the reaction mechanism of auto-phosphorylation and substitution, which activate or deactivate CIPKs. Several biochemical studies on CIPKs such as CIPK1, CIPK3, CIPK8, CIPK9, CIPK11, CIPK14, CIPK20, CIPK23, and CIPK24 have enhanced our understanding related to biochemical properties required for performing the phosphorylation reaction [3, 4, 6, 7, 16, 23, 26]. Unlike other kinases, most of the CIPKs preferred Mn^{2+} to Mg^{2+} as a divalent cation to coordinate the phosphate groups of the nucleotide triphosphate substrate [6, 8, 23].

G.K. Pandey et al., *Global Comparative Analysis of CBL–CIPK Gene Families in Plants*, SpringerBriefs in Plant Science, DOI 10.1007/978-3-319-09078-8_5

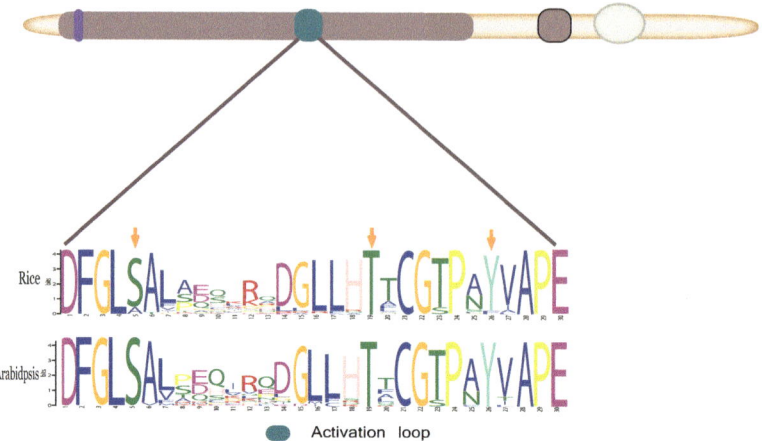

Fig. 5.1 Comparison of sequences of CIPKs activation loop in Arabidopsis and rice. Schematic diagram of highly conserved domain structure for CIPK showing an alignment of Arabidopsis and rice CIPKs activation loop. The three conserved residues in all members of CIPK gene family are shown by *arrow* on Ser(S), Thr(T), and Tyr(Y)

5.2 Mutagenesis of CIPKs to Generate Hyperactive Kinase or Dead Kinase

As typical characteristic of serine/threonine kinases, the catalytic domain of CIPK has ATP-binding loop and activation segment with DFG and APE motif (Fig. 5.1) [9]. The ATP binding is the site of phosphate binding at the N-terminal of the kinase. Mutating a conserved lysine (K) to asparagine (N) at the position 40 (which may vary in different CIPKs) can make the kinase dead by impairing its phosphate-binding capacity [7, 16]. The activation loop is located between the conserved DFG and APE residues in the catalytic kinase domain serine (S), threonine (T), and tyrosine (Y) residues shown by arrow in Fig. 5.1 and is almost conserved among all members of CIPK gene family in rice and Arabidopsis. By site-directed mutagenesis of the (T), (S), and (Y) to aspartate (D), a constitutively active kinase can be obtained, is one approach, whereas the other approach involves the removal of auto-inhibitory regulatory motif such as FISL/NAF domain [9, 20, 21], indicating that these residues are the sites of CIPK phosphorylation by other kinases *in vivo* to achieve active status. The removal of C-terminal auto-inhibitory segment of CIPK also generates a constitutively active kinase [9, 21]. Mutation of T to D and combining it with the removal of the regulatory segment acts synergistically to generate a hyperactive kinase.

Based on biochemical analysis, it is assumed that the N-terminal catalytic kinase domain of CIPKs is masked or inhibited by C-terminal regulatory domain and this intramolecular interaction blocks the substrates accessing into the kinase domain [8, 9]. Ca^{2+}-bound CBLs interact with C-terminal regulatory region, specifically the NAF/FISL motif and hence relieve the auto-inhibition exerted and hence release the kinase catalytic domain for access to its substrate [8]. Moreover, there are additional controls speculated on CIPK kinase activity through phosphorylation by upstream kinases in the activation loop along with auto-phosphorylation of key conserved residues to fine-tune the kinase activity [6, 9].

5.3 Function of Auto-phosphorylation in CIPKs

The introduction of a phosphate in the conserved residues of activation loop might result in ionic interactions that might be critical to enhance enzyme activity of kinases [15]. Besides substrate phosphorylation, CIPKs also undergo auto-phosphorylation in in vitro conditions. Recombinant CIPK24/SOS2 protein produced in bacteria exhibited no substrate phosphorylation activity in the absence of CBL4/SOS3, although it showed auto-phosphorylation activity under in vitro conditions [10]. In the presence of calcium, CBL4/SOS3 activated the substrate phosphorylation activity of CIPK24/SOS2 [10]. One of the auto-phosphorylated sites of CIPK24/SOS2 was found out to be Ser228. Mutation of Ser228 to Ala resulted in decreased auto-phosphorylation activity and substrate phosphorylation by CIPK24/SOS2. Moreover, genetic analysis, the mutation of Ser228 to Ala partially disrupted the function of CIPK24/SOS2 in salt tolerance, is seen in complementation test in *cipk24/sos2* mutant in Arabidopsis [5]. Interestingly, the auto-phosphorylation at Thr168 and auto-phosphorylation at Ser228 cannot completely substitute the function of CIPK24/SOS2 in genetic assays [5].

5.4 Physiological Target/Substrate of CIPKs

In the calcium signaling mediated by CBLs and CIPKs, the signal is transduced downstream by phosphorylation of certain responders such as transporters/channels, transcription factors, and other structural or regulatory proteins. For most of the CIPKs, the physiological substrate has not yet been identified. In order to understand the fundamental of signal transduction mechanism mediated by CBL–CIPK module, an in-depth knowledge of endogenous targets or substrate, which can be phosphorylated by CIPKs, would be important to improve our understanding of stimulus-response-coupling in plant cell. Some of the physiological substrates identified in Arabidopsis for CIPKs are the well-known SOS components, i.e., salt overly sensitive involved in salt stress tolerance, SOS1 (Na^+/H^+ antiporter), identified one of the first substrate of CIPK24/SOS2 [6, 8, 22]. The complete dissection of calcium-mediated signaling pathway regulating the uptake of K^+ ion during low-potassium condition has identified CIPK23 as a crucial kinase phosphorylating and regulating a high-affinity K^+ channel (AKT1) in Arabidopsis [16, 26]. In addition to the regulation of K^+ ion uptake, CIPK23 is also involved in phosphorylation-based regulation of a nitrate transporter CHL or NRT1 at threonine 101 [13, 14]. CIPK11, after being activated by CBL2, phosphorylate the cytosolic part of AHA2 (plasma membrane H^+ ATPase) at serine 931, which deactivates the ATPase [4]. Apart from channels, transcription factors ERF7 and ABI5 are phosphorylated by CIPK15 and CIPK26, respectively [18, 24]. CIPK26 also phosphorylates the C-terminal of a plant NADPH oxidase, RBOHF [2]. The CBL4–CIPK6 module is involved in regulation of AKT2 in potassium nutrition without phosphorylation of AKT2 channel [12]. One of the interesting parameter recently discovered that CIPKs could also phosphorylate their upstream interacting component/target, i.e., CBLs and this is discussed in detail in the following section [11, 19].

5.5 Phosphorylation of CBL by Their Interacting CIPK

CBLs are responsible for decoding the 'calcium signature' generated during a particular signaling event in the plant cell and act as calcium sensor relay by interacting and activating CIPKs, which possess the phosphotransferase enzymatic activity. In addition to phosphorylation of downstream targets such as transporters/channels, CIPKs can also phosphorylate CBLs and first reported in sweet pea, where PsCIPK could phosphorylate PsCBL in in vitro condition [19]. Subsequently in the year 2009, it was established that CBL10 is phosphorylated by SOS2/CIPK24 [17]. This phosphorylation was induced by salt stress, occurs at the plasma membrane, stabilizes the SCaBP8/CBL10–CIPK24/SOS2 interaction, and enhances plasma membrane Na^+/H^+ exchange activity [17]. The detailed biochemical assays led to identify CIPK24/SOS2 phosphorylate Arabidopsis CBL10 at Ser-237 residue of C-terminal but could not phosphorylate CBL4/SOS3. Furthermore, comprehensive biochemical assays indicated that CBLs are phosphorylated at the PFPF motif at the C-terminal [3]. CBLs are phosphorylated at serine (S)/threonine (T) residues for various CBL–CIPK pairs such as CBL10–CIPK24, CBL2–CIPK11, CBL2–CIPK14, CBL1–CIPK23, and CBL9–CIPK23. Moreover, Du et al. [3] reported that the phosphorylation increased the interaction strength between CBLs and CIPKs. Kudla and co-workers [11] reported that this phosphorylation acted synergistically to increase substrate phosphorylation by CIPKs. Beside Arabidopsis, the phosphorylation of CBLs by CIPKs has also been reported in several other plant species. In tomato, CIPK6 interacts with CBL10 and regulates the kinase activity in Ca^{2+}-dependent manner [1]. Inhibitory effect of Ca^{2+} on CIPK activity in the presence of CBL was shown for the first time [25], in *Vicia faba*, where VfCIPK1 was activated by VfCBL1 and showed a higher activity in the absence of Ca^{2+} than in its presence [25]. Additionally, CBL phosphorylation by their interacting CIPK is a critical component for complex stability and CBL–CIPK signaling pathway [3, 11]. CBL phosphorylation is required for regulation of the plasma membrane H^+-ATPase, AHA2 by CBL2–CIPK11 complex [3].

References

1. de la Torre F, Gutierrez-Beltran E, Pareja-Jaime Y, Chakravarthy S, Martin GB, Del Pozo O (2013) The tomato calcium sensor cbl10 and its interacting protein kinase cipk6 define a signaling pathway in plant immunity. Plant Cell 25:2748–2764
2. Drerup MM, Schlucking K, Hashimoto K, Manishankar P, Steinhorst L, Kuchitsu K, Kudla J (2013) The calcineurin B-like calcium sensors CBL1 and CBL9 together with their interacting protein kinase CIPK26 regulate the Arabidopsis NADPH oxidase RBOHF. Mol Plant 6:559–569
3. Du W, Lin H, Chen S, Wu Y, Zhang J, Fuglsang AT, Palmgren MG, Wu W, Guo Y (2011) Phosphorylation of SOS3-like calcium-binding proteins by their interacting SOS2-like protein kinases is a common regulatory mechanism in Arabidopsis. Plant Physiol 156:2235–2243
4. Fuglsang AT, Guo Y, Cuin TA, Qiu Q, Song C, Kristiansen KA, Bych K, Schulz A, Shabala S, Schumaker KS, Palmgren MG, Zhu JK (2007) Arabidopsis protein kinase PKS5 inhibits the plasma membrane H^+-ATPase by preventing interaction with 14-3-3 protein. Plant Cell 19:1617–1634
5. Fujii H, Zhu JK (2009) An autophosphorylation site of the protein kinase SOS2 is important for salt tolerance in Arabidopsis. Mol Plant 2:183–190
6. Gong D, Gong Z, Guo Y, Chen X, Zhu JK (2002) Biochemical and functional characterization of PKS11, a novel Arabidopsis protein kinase. J Biol Chem 277:28340–28350
7. Gong D, Guo Y, Jagendorf AT, Zhu JK (2002) Biochemical characterization of the Arabidopsis protein kinase SOS2 that functions in salt tolerance. Plant Physiol 130:256–264
8. Gong D, Guo Y, Schumaker KS, Zhu JK (2004) The SOS3 family of calcium sensors and SOS2 family of protein kinases in Arabidopsis. Plant Physiol 134:919–926
9. Guo Y, Halfter U, Ishitani M, Zhu JK (2001) Molecular characterization of functional domains in the protein kinase SOS2 that is required for plant salt tolerance. Plant Cell 13:1383–1400
10. Halfter U, Ishitani M, Zhu JK (2000) The Arabidopsis SOS2 protein kinase physically interacts with and is activated by the calcium-binding protein SOS3. Proc Natl Acad Sci USA 97:3735–3740
11. Hashimoto K, Eckert C, Anschutz U, Scholz M, Held K, Waadt R, Reyer A, Hippler M, Becker D, Kudla J (2012) Phosphorylation of calcineurin B-like (CBL) calcium sensor proteins by their CBL-interacting protein kinases (CIPKs) is required for full activity of CBL–CIPK complexes toward their target proteins. J Biol Chem 287:7956–7968
12. Held K, Pascaud F, Eckert C, Gajdanowicz P, Hashimoto K, Corratge-Faillie C, Offenborn JN, Lacombe B, Dreyer I, Thibaud JB, Kudla J (2011) Calcium-dependent modulation and plasma membrane targeting of the AKT2 potassium channel by the CBL4/CIPK6 calcium sensor/protein kinase complex. Cell Res 21:1116–1130
13. Ho CH, Lin SH, Hu HC, Tsay YF (2009) CHL1 functions as a nitrate sensor in plants. Cell 138:1184–1194
14. Ho CH, Tsay YF (2010) Nitrate, ammonium, and potassium sensing and signaling. Curr Opin Plant Biol 13:604–610
15. Johnson LN, Noble ME, Owen DJ (1996) Active and inactive protein kinases: structural basis for regulation. Cell 85:149–158
16. Li L, Kim BG, Cheong YH, Pandey GK, Luan S (2006) A Ca^{2+} signaling pathway regulates a K^+ channel for low-K response in Arabidopsis. Proc Natl Acad Sci USA 103:12625–12630
17. Lin H, Yang Y, Quan R, Mendoza I, Wu Y, Du W, Zhao S, Schumaker KS, Pardo JM, Guo Y (2009) Phosphorylation of SOS3-LIKE CALCIUM BINDING PROTEIN8 by SOS2 protein kinase stabilizes their protein complex and regulates salt tolerance in Arabidopsis. Plant Cell 21:1607–1619
18. Lyzenga WJ, Liu H, Schofield A, Muise-Hennessey A, Stone SL (2013) Arabidopsis CIPK26 interacts with KEG, components of the ABA signalling network and is degraded by the ubiquitin-proteasome system. J Exp Bot 64:2779–2791

19. Mahajan S, Sopory SK, Tuteja N (2006) Cloning and characterization of CBL–CIPK signalling components from a legume (*Pisum sativum*). FEBS J 273:907–925
20. Pandey GK (2008) Emergence of a novel calcium signaling pathway in plants: CBL–CIPK signaling network. Physiol Mol Biol Plants 14:51–68
21. Qiu QS, Guo Y, Dietrich MA, Schumaker KS, Zhu JK (2002) Regulation of SOS1, a plasma membrane Na^+/H^+ exchanger in *Arabidopsis thaliana*, by SOS2 and SOS3. Proc Natl Acad Sci USA 99:8436–8441
22. Quintero FJ, Ohta M, Shi H, Zhu JK, Pardo JM (2002) Reconstitution in yeast of the Arabidopsis SOS signaling pathway for Na^+ homeostasis. Proc Natl Acad Sci USA 99:9061–9066
23. Shi J, Kim KN, Ritz O, Albrecht V, Gupta R, Harter K, Luan S, Kudla J (1999) Novel protein kinases associated with calcineurin B-like calcium sensors in Arabidopsis. Plant Cell 11:2393–2405
24. Song CP, Agarwal M, Ohta M, Guo Y, Halfter U, Wang P, Zhu JK (2005) Role of an Arabidopsis AP2/EREBP-type transcriptional repressor in abscisic acid and drought stress responses. Plant Cell 17:2384–2396
25. Tominaga M, Harada A, Kinoshita T, Shimazaki K (2010) Biochemical characterization of calcineurin B-like-interacting protein kinase in Vicia guard cells. Plant Cell Physiol 51:408–421
26. Xu J, Li HD, Chen LQ, Wang Y, Liu LL, He L, Wu WH (2006) A protein kinase, interacting with two calcineurin B-like proteins, regulates K^+ transporter AKT1 in Arabidopsis. Cell 125:1347–1360

Chapter 6
Protein Interactions

Abstract A large number of CBLs and CIPKs have been identified in several plant species, which suggest a complex web of potential interaction between them. Preferential complex formation of individual CBLs with defined subsets of CIPKs contribute in generating specificity in signaling network. CIPKs transduce the signal further downstream in the signaling pathway. In this context, CIPKs have been found to interact with diverse and broad range of protein targets.

Keywords Interactions · Targets · Phosphatases · Transporters/channels · Transcription factors · Enzymes · Regulation · CBL · CIPK

6.1 Introduction

Immediately after the discovery of CBLs in Arabidopsis, the first interacting partners identified were CIPKs. As mentioned earlier both CBLs and CIPKs exist as multiple members in higher plants. With a large number of proteins for CBLs and CIPKs family, a complex network of CBL–CIPK interacting module is speculated. In Arabidopsis, there are 10 CBLs and 26 CIPKs so in principle; there are approximately 260 possible interacting modules, which represent a large diversification of the signaling cascades. However, one CBL can interact with a specific CIPK, leading to generate specificity in the signaling pathways and at the same time interact with multiple CIPKs for cross talk or overlapping in signaling pathways. Two approaches were widely used to analyze the interaction between CBLs and CIPKs, including yeast two-hybrid and bimolecular fluorescence complementation (BiFC).

© The Author(s) 2014   45
G.K. Pandey et al., *Global Comparative Analysis of CBL–CIPK Gene Families in Plants*, SpringerBriefs in Plant Science, DOI 10.1007/978-3-319-09078-8_6

6.2 Various Interactors of CBLs

6.2.1 CIPKs

In Arabidopsis, CBL–CIPK interactome has been well established and many of the CBL–CIPK modules are responsible for generating specificity and diversity in signaling networks [1, 17, 21]. As mentioned earlier that specific domains present on both CBLs and CIPKs are responsible for interaction between these components, however, molecular mechanism behind this interaction is still unknown. Many of these interactions were shown in heterologous system such as in yeast and also in different plant species. It is still unclear how and what factors are exactly responsible for specific overlapping interaction between CBLs and CIPKs in the physiological condition.

6.2.2 Others

With extensive research in the field of CBLs in Arabidopsis and other plant species, the CBLs are generally known to be the only interactors of CIPKs. But recently, some new reports have emerged showing novel interacting partners of CBLs beside CIPKs. In Arabidopsis, CBL1 interacts with voltage-dependent anion channel 1 (VDAC1) [14]. Several sugar-responsive and GA biosynthetic genes were altered in the *CBL1* mutant showing a possible existence of genetic interaction [15]. CBL1 protein physically interacted with AKINβ1, the regulatory β subunit of the SnRK1 complex, which has a central role in sugar signaling [15]. In another case, CBL3 interacts with AtMTAN (5′-methylthioadenosine nucleosidase) [19]. In-vitro and in-vivo analysis verified that CBL3 and AtMTAN physically associated only in the presence of Ca^{2+}. Ca^{2+}-bound CBL3 inhibits MTAN activity [19]. Similarly, AtCBL10 interacted with AKT1 directly without binding to any AtCIPKs and regulate K^+ nutrition and homeostasis [23]. There are evidences for a direct connection between CBL2/CBL3 and V-ATPase activity, but exact mechanism is still unknown [27].

6.3 Various Targets of CIPKs

In the calcium signaling pathway, CBLs sense the calcium signature and interact with CIPKs to transduce the signal downstream. Here, CBLs act as sensor relay and CIPKs are the effector proteins, and has the catalytic enzymatic activity to transfer a phosphate group covalently to its interacting partner and hence transduce the signal further downstream in the signaling pathway. In this context, CIPKs have been found to interact with diverse and broad range of protein targets as depicted in the Fig. 6.1.

Fig. 6.1 Diverse targets
of CIPKs. CIPKs interact
and target a large number
of interactors such as
phosphatases, transporters/
channels, transcription
factors, and enzymes

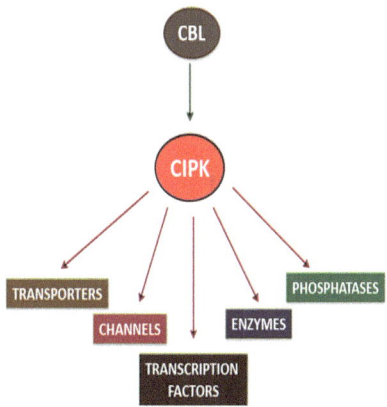

6.3.1 Phosphatases

In Arabidopsis, the coordinated action of a family of ten calcium sensors (CBLs),
twenty-six protein kinases (CIPKs), and phosphatases produces transient phosphorylation–dephosphorylation reaction and the activation of several signaling pathways
in response to environmental stresses. The major category of phosphatases interacting with CIPKs belongs to PP2C class as shown in Arabidopsis. The interaction
between CIPKs and PP2C phosphatases is specific in nature and the domain responsible for this interaction has been mapped out to be protein phosphatase interaction
(PPI) and kinase interaction motif (KIM), respectively [20]. At molecular level, it
can be said that the PP2C binds with the PPI domain of CIPK where it may control the autophosphorylation of the kinase or the phosphorylation of the substrate by
the kinase [20]. It may also be possible that CIPK act as a scaffold protein, carry-
ing PP2C to the target where both of them modulate the target [12]. Another model
proposed by Lan et al. [11] predicted to have more complexity where the PP2C can
bind to a domain in the catalytic region of CIPK and CBL binds to the PP2C to
modulate its activity [11]. The CBL–CIPK–PP2C yielded a new paradigm in understanding mechanistic regulation of calcium signaling modules *in-planta*. CIPK24/
SOS2 and CIPK15/PKS3 (SOS2-like protein kinase) interacted specifically with
abscisic acid (ABA)-insensitive2 (ABI2) phosphatase [7, 20].

6.3.2 Transporters/Channels

By genetic interaction analysis of salt overly sensitive (SOS) in salt-stress tolerance pathway, Zhu and co-workers first identified a transporter SOS1 also
known as Na^+/H^+-antiporter present on plasma membrane [10, 22, 24, 25].
SOS3/CBL4–SOS2/CIPK24 complex interact and phosphorylate SOS1, Na^+/H^+-
antiporter, on the plasma membrane and regulate salt tolerance [24]. Several

other transporters/channels such as AHA (H^+ ATPases) and CAX1 (cation/H^+-exchangers) were also found to be interacting with some of the CIPKs biochemically or genetically in regulating the transport of ions across plasma membrane or endomembrane [4, 6]. During K^+ uptake and nutrition, CBL1/9–CIPK23 pair was found to interact and phosphorylate a high affinity K^+ channel (AKT1) in root plasma membrane [13, 29]. Moreover, CBL4–CIPK6 complex was found to interact and target AKT2 to plasma membrane and is involved in K^+ homeostasis [8]. Similarly, CBL1/9–CIPK23 complex was also reported to interact and phosphorylate a nitrate transporter, NRT1.1 or CHL1 and regulated nitrate sensing and uptake [9]. Furthermore, formation of CBL–CIPK–PP2C complexes was also shown to be involved in fine-tuning of stress management, ion uptake, and homeostasis [24].

6.3.3 Transcription Factors

CIPK15/PKS3 interacts and phosphorylates ERF7 and negatively regulates ABA signaling [26], CIPK26 interact transcription factor such as ABI5, which are key components of ABA signaling pathways and phosphorylates ABI5 in vitro [18]. Consistent with a role in ABA signaling, overexpression of CIPK26 increased the sensitivity of germinating seeds to the inhibitory effects of ABA. In tomato (*Solanum lycopersicum* L.), in vitro pull-down assay of the recombinant SlCIPK2 confirmed interaction with stress-responsive transcription factors, SlERF7, SlCBF1, and SlAREB1 [30]. Different isoforms of CIPK3 interact and might be involved in phosphorylation-based regulation of ABR1(ABA repressor1) in transducing the ABA signaling pathway (Sanyal and Pandey, unpublished data).

6.3.4 Enzymes

In vivo and in vitro data indicates that CIPK26 after interacting with the calcium sensors CBL1 or CBL9 enhances the activity of the NADPH oxidase RBOHF via phosphorylation [5]. CIPK26 phosphorylates RBOHF in vitro and co-expression of either CBL1 or CBL9 with CIPK26 strongly enhances ROS production by RBOHF in HEK293T cells. Together, these findings identify a direct connection between CBL–CIPK-mediated Ca^{2+} and ROS signaling in plants. And this also provides evidence for a synergistic activation of NADPH oxidase RBOHF by direct Ca^{2+}-binding to its EF-hands and Ca^{2+}-induced phosphorylation by CBL1/9–CIPK26 complexes [5]. CIPK26 has also been shown to interact with RING-type E3 ligase, 'Keep on Going' (KEG)-interacting protein [18]. In vitro pull-down and *in-planta* bimolecular fluorescence complementation assays confirmed the interactions between CIPK26 and KEG. KEG functions by mediating proteasomal degradation of CIPK26 and regulating ABA signaling [18].

6.4 CBL–CIPK Complexes Regulate a Broad Range of Functions

As stated above diverse kinds of interactors are regulated by CBLs and CIPKs and modulation of these leads to activation or inhibition of various biological functions in plants [2, 17, 28]. In past few years, the role of CBL–CIPK has been greatly appreciated in regulating various biological processes such as abiotic stress, nutrition deficiency, biotic stress, development, and ROS signaling, which suggest a profound role of this calcium signaling machinery in the sustenance processes and life cycle of plants [3, 16, 21, 27].

References

1. Batistic O, Kudla J (2004) Integration and channeling of calcium signaling through the CBL calcium sensor/CIPK protein kinase network. Planta 219:915–924
2. Batistic O, Kudla J (2009) Plant calcineurin B-like proteins and their interacting protein kinases. Biochim Biophys Acta 1793:985–992
3. Batistic O, Rehers M, Akerman A, Schlucking K, Steinhorst L, Yalovsky S, Kudla J (2012) S-acylation-dependent association of the calcium sensor CBL2 with the vacuolar membrane is essential for proper abscisic acid responses. Cell Res 22(7):1155–1168
4. Cheng NH, Pittman JK, Zhu JK, Hirschi KD (2004) The protein kinase SOS2 activates the Arabidopsis H^+/Ca^{2+} antiporter CAX1 to integrate calcium transport and salt tolerance. J Biol Chem 279:2922–2926
5. Drerup MM, Schlucking K, Hashimoto K, Manishankar P, Steinhorst L, Kuchitsu K, Kudla J (2013) The Calcineurin B-like calcium sensors CBL1 and CBL9 together with their interacting protein kinase CIPK26 regulate the Arabidopsis NADPH oxidase RBOHF. Mol Plant 6:559–569
6. Fuglsang AT, Guo Y, Cuin TA, Qiu Q, Song C, Kristiansen KA, Bych K, Schulz A, Shabala S, Schumaker KS, Palmgren MG, Zhu JK (2007) Arabidopsis protein kinase PKS5 inhibits the plasma membrane H^+-ATPase by preventing interaction with 14-3-3 protein. Plant Cell 19:1617–1634
7. Guo Y, Xiong L, Song CP, Gong D, Halfter U, Zhu JK (2002) A calcium sensor and its interacting protein kinase are global regulators of abscisic acid signaling in Arabidopsis. Dev Cell 3:233–244
8. Held K, Pascaud F, Eckert C, Gajdanowicz P, Hashimoto K, Corratge-Faillie C, Offenborn JN, Lacombe B, Dreyer I, Thibaud JB, Kudla J (2011) Calcium-dependent modulation and plasma membrane targeting of the AKT2 potassium channel by the CBL4/CIPK6 calcium sensor/protein kinase complex. Cell Res 21:1116–1130
9. Ho CH, Lin SH, Hu HC, Tsay YF (2009) CHL1 functions as a nitrate sensor in plants. Cell 138:1184–1194
10. Ishitani M, Liu J, Halfter U, Kim CS, Shi W, Zhu JK (2000) SOS3 function in plant salt tolerance requires N-myristoylation and calcium binding. Plant Cell 12:1667–1678
11. Lan WZ, Lee SC, Che YF, Jiang YQ, Luan S (2011) Mechanistic analysis of AKT1 regulation by the CBL–CIPK–PP2CA interactions. Mol Plant 4:527–536
12. Lee SC, Lan WZ, Kim BG, Li L, Cheong YH, Pandey GK, Lu G, Buchanan BB, Luan S (2007) A protein phosphorylation/dephosphorylation network regulates a plant potassium channel. Proc Natl Acad Sci USA 104:15959–15964
13. Li L, Kim BG, Cheong YH, Pandey GK, Luan S (2006) A Ca^{2+} signaling pathway regulates a K^+ channel for low-K response in Arabidopsis. Proc Natl Acad Sci USA 103:12625–12630

14. Li ZY, Xu ZS, Chen Y, He GY, Yang GX, Chen M, Li LC, Ma YZ (2013) A novel role for Arabidopsis CBL1 in affecting plant responses to glucose and gibberellin during germination and seedling development. PLoS ONE 8:e56412

15. Li ZY, Xu ZS, He GY, Yang GX, Chen M, Li LC, Ma Y (2013) The voltage-dependent anion channel 1 (AtVDAC1) negatively regulates plant cold responses during germination and seedling development in Arabidopsis and interacts with calcium sensor CBL1. Int J Mol Sci 14:701–713

16. Liu LL, Ren HM, Chen LQ, Wang Y, Wu WH (2013) A protein kinase, calcineurin B-like protein-interacting protein Kinase9, interacts with calcium sensor calcineurin B-like Protein3 and regulates potassium homeostasis under low-potassium stress in Arabidopsis. Plant Physiol 161:266–277

17. Luan S (2009) The CBL–CIPK network in plant calcium signaling. Trends Plant Sci 14:37–42

18. Lyzenga WJ, Liu H, Schofield A, Muise-Hennessey A, Stone SL (2013) Arabidopsis CIPK26 interacts with KEG, components of the ABA signalling network and is degraded by the ubiquitin-proteasome system. J Exp Bot 64:2779–2791

19. Oh SI, Park J, Yoon S, Kim Y, Park S, Ryu M, Nam MJ, Ok SH, Kim JK, Shin JS, Kim KN (2008) The Arabidopsis calcium sensor calcineurin B-like 3 inhibits the 5′-methylthioadenosine nucleosidase in a calcium-dependent manner. Plant Physiol 148:1883–1896

20. Ohta M, Guo Y, Halfter U, Zhu JK (2003) A novel domain in the protein kinase SOS2 mediates interaction with the protein phosphatase 2C ABI2. Proc Natl Acad Sci USA 100:11771–11776

21. Pandey GK (2008) Emergence of a novel calcium signaling pathway in plants: CBL–CIPK signaling network. Physiol Mol Biol Plants 14:51–68

22. Qiu QS, Guo Y, Dietrich MA, Schumaker KS, Zhu JK (2002) Regulation of SOS1, a plasma membrane Na^+/H^+ exchanger in *Arabidopsis thaliana*, by SOS2 and SOS3. Proc Natl Acad Sci USA 99:8436–8441

23. Ren XL, Qi GN, Feng HQ, Zhao S, Zhao SS, Wang Y, Wu WH (2013) Calcineurin B-like protein CBL10 directly interacts with AKT1 and modulates K^+ homeostasis in Arabidopsis. Plant J 74:258–266

24. Sanchez-Barrena MJ, Martinez-Ripoll M, Albert A (2013) Structural biology of a major signaling network that regulates plant abiotic stress: the CBL–CIPK mediated pathway. Int J Mol Sci 14:5734–5749

25. Shi H, Ishitani M, Kim C, Zhu JK (2000) The *Arabidopsis thaliana* salt tolerance gene SOS1 encodes a putative Na^+/H^+ antiporter. Proc Natl Acad Sci USA 97:6896–6901

26. Song CP, Agarwal M, Ohta M, Guo Y, Halfter U, Wang P, Zhu JK (2005) Role of an Arabidopsis AP2/EREBP-type transcriptional repressor in abscisic acid and drought stress responses. Plant Cell 17:2384--2396

27. Tang RJ, Liu H, Yang Y, Yang L, Gao XS, Garcia VJ, Luan S, Zhang HX (2012) Tonoplast calcium sensors CBL2 and CBL3 control plant growth and ion homeostasis through regulating V-ATPase activity in Arabidopsis. Cell Res 22:1650–1665

28. Weinl S, Kudla J (2009) The CBL–CIPK Ca^{2+}-decoding signaling network: function and perspectives. New Phytol 184:517–528

29. Xu J, Li HD, Chen LQ, Wang Y, Liu LL, He L, Wu WH (2006) A protein kinase, interacting with two calcineurin B-like proteins, regulates K^+ transporter AKT1 in Arabidopsis. Cell 125:1347–1360

30. Yuasa T, Ishibashi Y, Iwaya-Inoue M (2012) A flower specific calcineurin B-like molecule (CBL)-interacting protein kinase (CIPK) homolog in tomato cultivar micro-tom (*Solanum lycopersicum* L.). Am J Plant Sci 3:753–763

Chapter 7
Functional Role of CBL–CIPK in Nutrient Deficiency

Abstract In the field of CBL–CIPK signaling in plant, the majority of the work has been done in Arabidopsis. But with the identification of CBL–CIPK modules in several other plant species, a functional conservation of CBL–CIPK-mediated signaling networking has been observed. Here, we emphasize the processes where CBL–CIPK pair is involved to regulate the nutritional deficiency signaling cascades.

Keywords Function · Nutrient · Nitrate · Potassium deficiency · CBL · CIPK

7.1 Introduction

Plants are primarily dependent on soil for absorbing water and nutrients by roots. In the soil, nutrients availability highly affects the growth and development of the plants. In the poor-nutrient soil, plant senses the changes in nutrient concentration and then responds by activation of molecular machinery through signal transduction to survive nutrient-deprived conditions [37, 43, 49]. Based on several reports in the literature, CBL–CIPK module is primarily regulating the nitrate and potassium uptake and sensing [6, 24, 33, 44, 58, 59].

7.2 Nitrate Deficiency

Nitrogen is present in the atmosphere in abundant form. NO_3^- (nitrate) and NH_4^+ (ammonium), the primary nitrogenous nutrients absorbed by the plant roots from the soil, which act as building blocks of organic matter, cofactors, or as signaling molecules [7, 16, 17, 54]. After sensing the nitrogen availability in the soil, plant regulates nitrogen transport system. Nitrogen transport system consists of several components involved in the nitrogen acquisition, distribution, and signaling network [54]. Nitrogen nutrition and uptake physiology are very well-studied field

© The Author(s) 2014

G.K. Pandey et al., *Global Comparative Analysis of CBL–CIPK Gene Families in Plants*, SpringerBriefs in Plant Science, DOI 10.1007/978-3-319-09078-8_7

[1, 8, 13, 14, 24, 34, 40]. High-affinity transport system (HATS) and low-affinity transport system (LATS) are two important systems employed for uptake of nitrogen by roots [5, 15, 25, 50, 51].

Besides being important component of various biological molecules such as proteins and nucleic acids, nitrates in higher plants also act as a signaling molecule. Plants absorb nitrogen from soil either as nitrates or ammonia, utilizing two systems of nitrate uptake, a high-affinity system, and a low-affinity system. Low-affinity transporters are inducible under surplus nitrate condition, whereas high-affinity nitrate transporters are inducible under nitrate deficiency. NRT2.1 and NRT2.2 act in high-affinity uptake system [42]. NRT1.2 acts in low-affinity uptake system while NRT1.1 (formally called CHL1) is a dual affinity transporter for nitrate uptake [35, 41, 45, 57] (Fig. 7.1). CHL1 exhibits two phases of nitrate uptake, with a Km of 50 μM for the high-affinity phase and a Km of about 4 mM for the low-affinity phase [41, 42, 45].

CIPK8 is rapidly induced by nitrate and is a positive regulator of the primary nitrate response [27]. CIPK8 regulates the low-affinity phase of the primary nitrate response by targeting CHL1 (Fig. 7.2). However, a clear physical interaction and phosphorylation of CHL1 by CIPK8 have not been elucidated yet. CIPK23 is also involved in the primary nitrate response [24]. CIPK23 interacts and phosphorylates CHL1 [24]. In the absence of phosphorylation by CIPK23, the high-affinity transport activity is inhibited in CHL1 (Fig. 7.2). Moreover, CHL1 acts as a nitrate sensor and can detect a wide range of concentration changes and lead to different levels of responses in the plants [24]. The nitrate sensing and uptake by CHL1 gave rise to the term called 'transceptor' for a transporter and receptor [52]. This mechanism might also serve as a prototype to understand how nutrient concentration changes are detected in other organisms (Fig. 7.2).

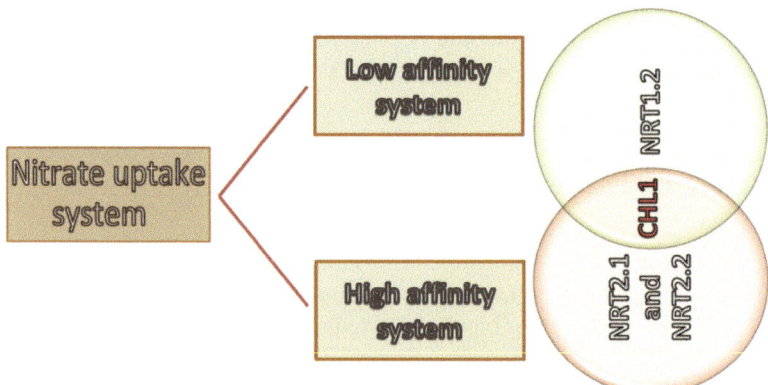

Fig. 7.1 CHL1—a dual affinity nitrate transporter. Nitrate uptake system in plants is divided into two modes, high-affinity system and low-affinity system in which NRT2.1 and NRT2.2 works in high-affinity system and NRT1.2 in low-affinity system. CHL1 (NRT1.1) is considered to be involved in both high- and low-affinity nitrate uptake system

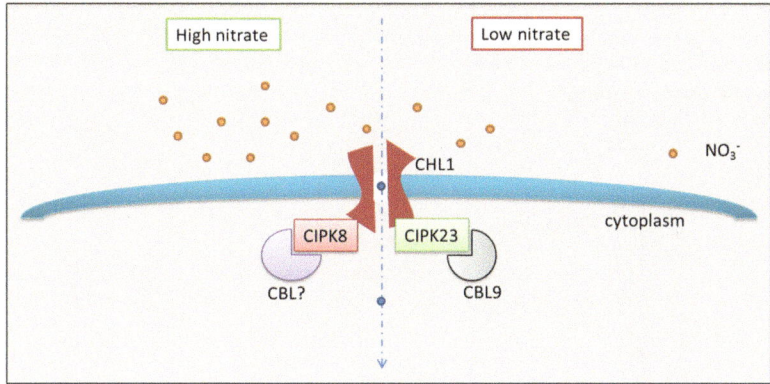

Fig. 7.2 Involvement of CBL–CIPK in nitrate signaling pathway. CIPK8 regulates the low-affinity phase whereas CIPK23 regulates the high-affinity phase of nitrate uptake

7.3 Potassium Deficiency

Potassium is a plant macronutrient. It is present as inorganic cation in the cytosol [32]. K^+ deficiency leads to major developmental abnormalities, nitrogen and sugar imbalance, impaired photosynthesis, and long-distance transport [39]. Plants have developed adaptive measures to cope K^+ deficiency, involving various signaling modules [4, 38, 53]. Recently, CBL–CIPK module has been found to regulate the K^+ signaling and uptake pathways in Arabidopsis [23, 33, 44, 58].

7.3.1 CBL–CIPK23–AKT1

In a breakthrough research in 2006, two major groups (Luan and co-worker and Wu and colleagues) working in the field of K^+ uptake and nutrition signaling have identified the role of CBL–CIPK in regulating the signaling pathway in response to K^+ deprivation [33, 58]. By detailed biochemical, electrophysiological, and genetic-based approaches, CBL1, CBL9, and CIPK23 were identified to be involved in a Ca^{2+}-dependent regulation of AKT1, a high-affinity voltage-gated potassium channel in Arabidopsis root cell functional during low-K conditions (Fig. 7.3).

By a forward genetic screen, Xu et al. [58] identified the low-potassium sensitive (*lks*) mutant, exhibiting severe sensitivity as compared with wild type, based on the chlorosis of the seedling in low-potassium containing media. Upon cloning of this *lks* mutant by map-based cloning led to identify a Ser/Thr protein kinase, CIPK23. By yeast two-hybrid analysis, the upstream and downstream targets of CIPK23 were identified as CBL1 and 9, calcium sensors and AKT1, a high-affinity voltage-dependent potassium channel, respectively [58]. In genetic analysis of *lks* (*cipk23*), *cbl1*, *cbl9*, and *akt1* mutants, chlorosis/bleaching of leaves in

Fig. 7.3 Involvement of CBL–CIPK in potassium signaling pathways. This represents the three signaling pathways in Arabidopsis for low-K$^+$ tolerance and adaptation

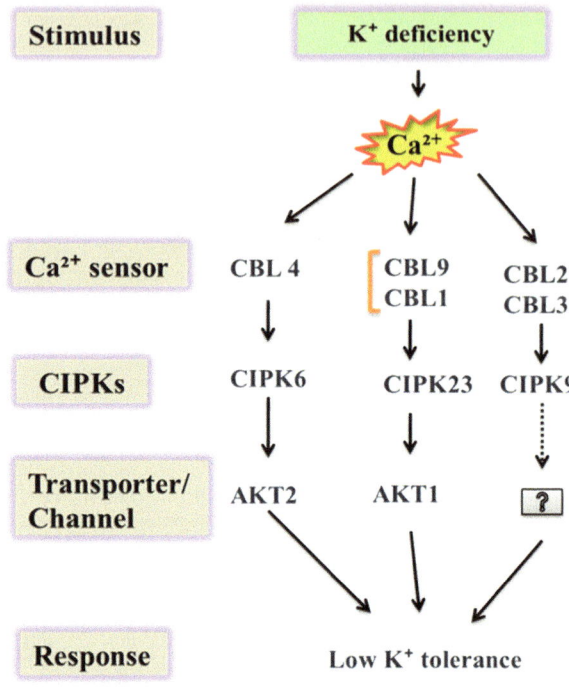

low-potassium containing media was observed for *lks* and *akt1* mutants; however, single mutants of *cbl1* and *cbl9* did not show any chlorosis/bleaching of leaves on low-potassium containing media, suggesting they act redundantly. Interestingly, the double mutants of *cbl1cbl9* showed a similar chlorosis/bleaching or hypersensitive phenotype to low-potassium media. By gain-of-function approach of *CIPK23*, *CBL1*, and *CBL9*, overexpressing transgenic plants showed better growth or tolerance on low-potassium containing media.

Concurrently, Luan group also identified CIPK23 as a major player regulating potassium nutrition by gene expression and systematic reverse genetic analysis [6, 33]. During the systematic gene expression of all the *CBLs* and *CIPKs*, the transcripts of *CIPK23* were found to be up-regulated under potassium deprivation conditions, and this inducible nature of *CIPK23* under K$^+$ deficiency condition led them to perform detailed functional investigation under low-potassium conditions [6, 33]. By using reverse genetic approach, T-DNA-inserted mutant alleles of CIPK23 were analyzed under potassium-deficient growth media, *cipk23* mutant alleles showed sensitive nature of seedling growth (mostly root growth) [6], unlike the chlorosis/bleaching of leaves-based phenotype observed by Xu et al. [58]. Here, the differences in phenotype observed by Wu and Luan groups are because of differences in the methodologies to assess the phenotype of *cipk23* mutant alleles on K$^+$-deficient conditions [52].

Similarly, Cheong et al. [6] identified the upstream calcium sensors CBL1 and CBL9, synergistically required to regulate CIPK23-mediated low-potassium nutrition. The functional analysis of AKT1 channel activity and its regulation were performed in *Xenopus* oocyte system by electrophysiological two-electrode voltage clamp (TEV) analytical tools. The inward potassium channel activity of AKT1 was fully reconstituted when CBL1 or CBL9–CIPK23 module was co-injected along with AKT1 [33, 58] compared to when any of the components of the module were missing. By in vitro phosphorylation analysis, Li et al. [33] confirmed that CIPK23 phosphorylate AKT1 channels at the C-terminal region. Similarly, the channel activity was shown to be regulated by phosphorylation in vivo in oocyte where only the active CIPK23 and not inactive (ATP-binding site mutant CIPK23 variant, CIPK23T60 N) could activate the AKT1 channel [33]. In addition to functional reconstitution of CBL1/9–CIPK23–AKT1 pathway in heterologous system, i.e., *Xenopus* oocyte, the in vivo role of this pathway was analyzed in the loss-of-function mutant *cipk23, cbl1cbl9* double mutant and wild type by patch-clamp methodology in root hair cell to detect the inward-rectifying current [33, 58]. The results of this analysis also indicated that AKT1 channel activity was reduced in both *cipk23* and *cbl1cbl9* double mutants as compared with wild type, which further confirmed the functional role of CBL1/CBL9–CIPK23 module in regulating AKT1 *in planta* [33, 58].

7.3.2 CBL2/3–CIPK9-Unknown Target

After the discovery of several K^+ channels and transporters in Arabidopsis and other plants, one of the important questions before plants physiologist is how this transport is regulated under different conditions. To address this question, several groups were trying to identify the upstream regulatory components such as kinases or phosphatases, which might be fine-tuning the uptake and homeostasis of K^+ in the plant cell under different conditions. After the discovery of calcium-mediated CBL1/9–CIPK23 regulating a shaker family high-affinity K^+ channel, AKT1, the avenues for understanding of molecular mechanisms of K^+ uptake and signaling have started to unveil the regulatory aspect. But it is quite obvious that the trait of K^+ uptake and nutrition signaling will be regulated by many other signaling components and hence multiple transporters/channels and candidates might be participating in sensing and relaying the signals through multiple signaling pathways. Also several of these transporters/channels might be regulated both transcriptionally and post-translationally during different conditions [38]. Therefore, it is important to investigate the other candidates involved in regulating the potassium nutrition and signaling pathways.

This quest for identification of alternate signaling pathways and channels/transporters involved in K^+ uptake, distribution, and homeostasis was systematically undertaken by Luan and co-workers where they have identified another member of CIPK family, *CIPK9* whose transcript level accumulated tremendously under K$^+$-deficient condition in Arabidopsis [44]. The functional role of this kinase was investigated by reverse genetic approach by isolating the T-DNA inserted null mutant alleles and assessment of phenotype on K$^+$-deficient media. The germination-based growth of these *cipk9–1* and *cipk9–2* mutant alleles exhibited reduced root and seedling growth [44]. This growth impairment phenotype was specifically observed under potassium deficiency at very low concentration of K$^+$ in the growth media. Surprisingly, the growth sensitivity phenotype of *cipk9* was mostly seen in supra-micromolar range, i.e., less than 20 μM K$^+$ compared with sub-micromolar range, i.e., 100 μM K$^+$ for *cipk23* mutant alleles [6, 44]. One of the important difference observed in *cipk9* mutant when compared with *cipk23* mutant was, the former showed no significant difference in K$^+$ uptake and content, in both potassium-sufficient (20 mM K$^+$) and potassium-deficient (20 μM K$^+$) conditions [44]. Moreover, by yeast two-hybrid assays, Pandey et al. [44], did not find the interaction between CIPK9 and C-terminal of AKT1. Till now, no downstream target of CIPK9 is discovered but authors have made a speculation that CIPK9 might be involved in regulating an alternative pathway other than CBL1/9–CIPK23–AKT1 [33, 44, 58].

Recently, Wu and colleagues have also investigated the role of CIPK9 in K$^+$ nutrition and homeostasis and identified two calcium sensors, CBL2 and CBL3, interacting with CIPK9 and targeting it to vacuolar membrane (tonoplast) [36]. However, the exact target of CIPK9 on vacuolar membrane has not been discovered yet, which might be involved in regulating the K$^+$ homeostasis across the vacuolar membrane (Fig. 7.3). In order to identify the downstream targets of CIPK9, we have done extensive yeast two-hybrid screening and identified several putative targets of CIPK9 (G.K. Pandey laboratory, unpublished data). Primarily, we categorized these CIPK9 targets into signaling components, transporters/channels, regulatory proteins involved in mitochondrial and vacuolar function, and signaling pathways to regulate the K$^+$ and oxidative stress response in the cell (G.K. Pandey laboratory, unpublished data). The detailed functional characterization of these candidates by genetic and biochemical tools will enable to understand the mechanistic role played by CIPK9 in regulating K$^+$ homeostasis and other-related signaling networks.

7.3.3 Other CBL–CIPKs Regulating AKT1

At the onset of discovery of regulation of K$^+$ uptake and nutrition by calcium-mediated CBL–CIPK pathway of AKT1 by post-translational modification at plasma membrane of root cell has opened up new frontiers in the field of K$^+$ nutrition [22, 33, 58]. At the same time, many questions were raised regarding the mechanistic interplay of CBL–CIPK modules regulating K$^+$ channel, AKT1. One of the important observations made by Li et al. [33], in their *in planta* recording of K$^+$ channel activity in *cbl1cbl9* double or *cipk23* mutant, was not abolished completely, which could be possible because of other CBL–CIPK complexes might also be involved in regulating the AKT1 channels activity. In order to address this question, Lee et al. [31] performed extensive yeast two-hybrid interaction assays between 10 CBLs and 26 CIPKs and the C-terminal of AKT1 and 26 CIPKs to determine interacting partners of AKT1. In their comprehensive yeast two-hybrid studies, two more CIPKs, CIPK6, and CIPK16 were also found to interact with AKT1 in addition to CIPK23. Similarly, two more CBLs, CBL2 and CBL3 in addition to CBL1 and CBL9, reported in earlier studies [33, 58] were found to interact with all three CIPKs including CIPK6, CIPK16, and CIPK23 [31]. In total, 4 CBLs and 3 CIPKs were found to be interacting with CIPK23 and AKT1, respectively, forming a multivalent interacting network [31]. As observed by Lee et al. [31], a multiple interaction networks of CBL–CIPK and AKT1 might be operating in the cell where multiple CBLs and CIPKs might be cooperatively regulating AKT1 activity.

Since deciphering the role of multiple CBLs and CIPKs in regulation of AKT1 channel activity in planta is a difficult task, therefore, a heterologous electrophysiological system, *Xenopus* oocytes, was employed for studying the effect of multiple CBLs and CIPKs on AKT1 channel activity [31]. A comprehensive electrophysiological analysis of these multiple CBLs and CIPKs on regulation of AKT1 channel activity identified a differential mode where a specific CBL and CIPK activate the channel differently than the other. For example, interaction of CBL1–CIPK23 with AKT1 produces the strongest channel activity whereas interaction combination of CBL2, 3, or 9 with CIPK6 or CIPK16 with AKT1 produces weaker channel activity [31] and hence indicating the degree of overlap and complexity in the cellular regulation by CBL–CIPK signaling network.

AKT1 can interact with multiple CIPKs, and the specificity of this interaction is because of ankyrin repeat domain at the C-terminus of AKT1 [31]. Based on the biochemical analysis, the ankyrin repeat domain of AKT1 was identified as a docking site, interacting with the kinase domain of the CIPKs, thereby determining the specificity of interaction of AKT1 channel protein with different CIPKs. In addition, Lee et al. [31] also discovered a novel protein phosphatase belonging to PP2C family known as AIP1, as interacting partner of AKT1 as well as CIPK23, implicated in dephosphorylating the AKT1 channel and hence reducing its inward-rectifying current in electrophysiological assays in *Xenopus* oocytes [31]. With the discovery of important phosphorylation–dephosphorylation switch of K^+ channel activity of AKT1, these findings shed a tremendous amount of light in the physiology of K^+ uptake and acquisition from root.

As mentioned above, AKT1 activity can be regulated by multiple CIPKs in addition to CIPK23, and one of the examples is CIPK6. Very recently, a detailed mechanistic study showed the involvement of CBL–CIPK–PP2C-type phosphatase in regulation of AKT1 channel by yeast two-hybrid and electrophysiological analysis [29]. The inhibitory effect of multiple A-type protein phosphatase 2C (PP2C) members such as AIP1, PP2CA, and AHG1 on CIPK6-mediated activation of AKT1 channels showed a complex regulatory role of these CBL–CIPK and PP2C in regulating the K^+ uptake and distribution [29]. By using extensive yeast two-hybrid and electrophysiological procedure, specific interaction of PP2CA was detected with CIPK6, which inactivates the AKT1 channel activity. However, several CBLs have positive effect in regulating the AKT1 channel activity by relieving the auto-inhibitory effect of regulatory domain from catalytic kinase domain as well as interacting directly and inhibiting PP2Cs [29]. With these findings, a detailed mechanistic role of CBL–CIPK–PP2CA complex has been identified in regulation of AKT1 activity in K^+ uptake and distribution process in plant.

In an interesting finding by Hashimoto et al. [21], the role of CBL1 phosphorylation on influencing the activity of AKT1 by CBL1–CIPK23 module unraveled another layer of complexity in the CBL–CIPK-mediated regulation of K^+ channel activity. In their study, CBL1 was phosphorylated by CIPK23 at conserved serine residue in PFPF motif and this phosphorylation does not influence the stability, localization, or CIPK interaction of these calcium sensor proteins. However, the phosphorylated CBL1 enhances the AKT1 channel activity in electrophysiological assays in *Xenopus* oocyte system [21]. These findings where a sensor is phosphorylated by the effector component to exert another layer of control in fine-tuning the activity of their substrate, i.e., the K^+ channel, unearthed the complexity in regulation by signaling components of a physiological process.

7.3.4 CBL4–CIPK6–AKT2

A large number of genes responsible for uptake and transport of K^+ primarily are classified as transporters or carrier and channels [19, 20, 48, 53, 55]. The potassium transporters or carrier are potassium transporter/potassium uptake transporter/high-affinity potassium transporters (KT/KUP/HAK), which includes the transcriptionally, highly inducible, HAK5 under potassium deprivation condition [2, 20, 28, 46, 47, 56]. The major potassium channels in plants are comprised of three families based on the homologous members from animal system [30]. The first major family of potassium channel is Shaker-type as designated from the *Drosophila* potassium transporter 'Shaker,' which is voltage-gated ion channel. The most well-characterized shaker family members are AKT1 and KAT1 (both of these are inward-conducting K^+ channels) [3]. The other shaker family of channels extensively studied is GORK and SKOR, which are outward-conducting potassium channels [18, 26]. The second family of channels includes tandem-pore K^+ channels (TPK), which are voltage -independent and mostly present in the membrane of vacuoles [11, 12]. The third family of potassium channels is Kir channel from animals also known as two transmembrane domain inward-rectifying channels [30]. Once the detailed mechanistic regulation of AKT1 channel has been discovered in uptake of K^+ and nutrition signaling by calcium-mediated CBL–CIPK–PP2C components, several groups have investigated the regulation of other K^+ channels.

Interestingly, one of the reports by Held et al. [23] revealed a fascinating aspect of regulation of a K^+ channel by a CBL–CIPK module in a non-phosphorylation-dependent mode. By yeast two-hybrid analysis, they have shown the interaction between CBL4 and CIPK6 and also confirmed interaction between CIPK6 and AKT2. Moreover, by using the cell biological and electrophysiological analysis they showed, CBL4–CIPK6 complex interacts with C-terminal of AKT2 and functionally constitutes the activation of inward-rectifying current of AKT2 activity in *Xenopus* oocyte (Fig. 7.3). A new dimension has been opened up in the regulation of AKT2 channel for efficient translocation from ER to plasma membrane by CBL4–CIPK6 complex. In contrary to AKT1 channel regulation by kinase activity of CIPKs, i.e., phosphorylation mediated by CBLs–CIPKs complex [29, 31, 33, 58], the phosphorylation of AKT2 by CIPK6 is not required for the functional activation of AKT2. However, the interaction of kinase with the AKT2 C-terminal and its targeting from ER to plasma membrane are critical for proper regulation of AKT2 activity [23]. Moreover, the dual lipid modifications of CBL4 by myristoylation as well as palmitoylation are also crucial for the translocation of AKT2 channels to plasma membrane by CBL4–CIPK6 complex [23]. This study suggests existence of multiple K^+ channels trafficking to membrane where calcium-dependent targeting of channels or transporters to the plasma membrane can be efficiently mediated by CBL–CIPK ensemble. This efficient trafficking is achieved by interaction with CIPK6 kinase, which along with CBL4 act as scaffolding protein for proper translocation of the channels to plasma membrane rather than employing phosphorylation switch to regulate the activity of channels.

In conclusion, many unknown and exciting possibilities might be exerting the regulation of ion channels and transporters by a complex array of CBL–CIPK signaling pathway, which still needs to be explored.

7.3.5 CBL–CIPK Regulating K^+ Nutrition in Other Plants Species

Similar homologous components prevail in other plant species such as in grape (*Vitis vinifera*), VvCBL01 and VvCIPK04 were found to be homologs of Arabidopsis AtCBL1 and AtCIPK23 (Fig. 7.4) [9]. Arabidopsis AtAKT1 (Arabidopsis K^+ transporter 1) counterpart in *V. vinifera* VvK1.1 can be stimulated by co-expression of AtCIPK23–AtCBL1 in *Xenopus* oocytes [9]. Not only this, VvCIPK4–VvCBL1 and VvCIPK3–VvCBL2 pairs can stimulate another potassium channel VvK1.2 [9]. VvK1.2 acts as a voltage-gated inward-rectifying K^+ channel and remained electrically silent when expressed alone in *Xenopus* oocytes. VvK1.2 channel is specifically expressed in the berry and mediate rapid K^+ transport in the berry and contribute to the extensive reorganization of the translocation pathways and transport mechanisms that occur at veraison.

Interestingly, *Populus euphratica*, *PeCBL1* when expressed in the Arabidopsis *cbl1cbl9* mutant could build the resistance of sensitive phenotypes to low K^+ and display a salt-sensitive phenotype compared with the mutant [60]. The *PeCBL1* transgenic plant root exhibited a higher capacity to absorb K^+ after exposure

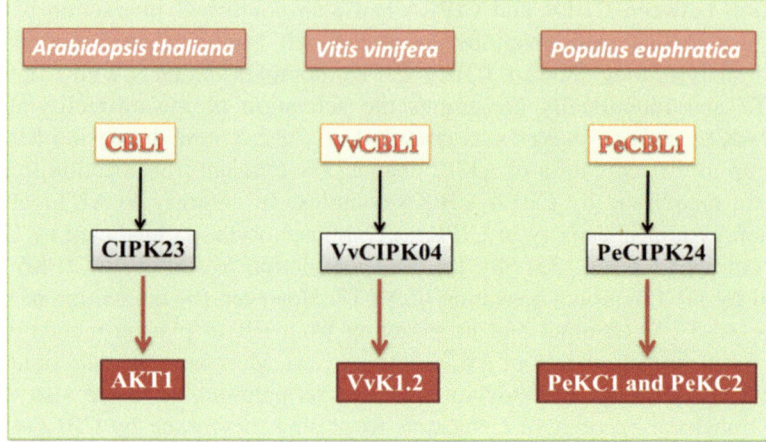

Fig. 7.4 Existence of similar components and pathways in different plant species. This figure represents parallel pathways in different genera of plants for regulating potassium uptake and homeostasis during K^+ deficiency

to low K^+ stress, and a lower capacity to discharge Na^+ after exposure to salt stress as compared with the *cbl1cbl9* mutant. PeCBL1 interacts with PeCIPK24, PeCIPK25, and PeCIPK26 to regulate Na^+/K^+ homeostasis in *P. euphratica,* which has a strong negative impact on fruit acidity. In *P. euphratica,* two homologs of shaker-like potassium channels were isolated and named PeKC1 and PeKC2. Over-expression of PeKC1 or PeKC2 in Arabidopsis *akt1* mutant rescues the mutant phenotype sensitive to low-K^+ stress [61]. In addition, PeKC1 and PeKC2 interact with PeCIPK24, a homolog of AtCIPK23 (Fig. 7.4) [10].

Based on the presence of similar homologous pathways, which are involved in regulation of K^+ transporters or channels, it can be stated that many of the physiological processes in plants do utilize highly conserved signaling components or networks in performing important and vital physiological processes.

References

1. Aguera E, Haba P, Fontes AG, Maldonado JM (1990) Nitrate and nitrite uptake and reduction in intact sunflower. Planta 182:149–154
2. Ahn SJ, Shin R, Schachtman DP (2004) Expression of *KT/KUP* Genes in Arabidopsis and the role of root hairs in K^+ uptake. Plant Physiol 134:1135–1145
3. Anderson JA, Huprikar SS, Kochian LV, Lucas WJ, Gaber RF (1992) Functional expression of a probable *Arabidopsis thaliana* potassium channel in Saccharomyces cerevisiae. Proc Natl Acad Sci USA 89:3736–3740
4. Armengaud P, Breitling R, Amtmann A (2004) The potassium-dependent transcriptome of Arabidopsis reveals a prominent role of jasmonic acid in nutrient signaling. Plant Physiol 136:2556–2576

5. Aslam M, Travis RL, Huffaker RC (1992) Comparative kinetics and reciprocal inhibition of nitrate and nitrite uptake in roots of uninduced and induced barley (*Hordeum uulgare* L.) seedlings. Plant Physiol 99:1124–1133

6. Cheong YH, Pandey GK, Grant JJ, Batistic O, Li L, Kim BG, Lee SC, Kudla J, Luan S (2007) Two calcineurin B-like calcium sensors, interacting with protein kinase CIPK23, regulate leaf transpiration and root potassium uptake in Arabidopsis. Plant J 52:223–239

7. Coruzzi GM, Bush DR (2001) Nitrogen and carbon nutrient and metabolite signaling in plants. Plant Physiol 125:61–64

8. Crawford NM, Glass ADM (1998) Molecular and physiological aspects of nitrate uptake in plants. Trends Plant Sci 3:389–395

9. Cuellar T, Azeem F, Andrianteranagna M, Pascaud F, Verdeil JL, Sentenac H, Zimmermann S, Gaillard I (2013) Potassium transport in developing fleshy fruits: the grapevine inward K^+ channel VvK1.2 is activated by CIPK–CBL complexes and induced in ripening berry flesh cells. Plant J 73:1006–1018

10. Cuellar T, Pascaud F, Verdeil JL, Torregrosa L, Adam-Blondon AF, Thibaud JB, Sentenac H, Gaillard I (2010) A grapevine Shaker inward K^+ channel activated by the calcineurin B-like calcium sensor 1-protein kinase CIPK23 network is expressed in grape berries under drought stress conditions. Plant J 61:58–69

11. Czempinski K, Frachisse JM, Maurel C, Barbier-Brygoo H, Muller-Rober B (2002) Vacuolar membrane localization of the Arabidopsis 'two-pore' K^+ channel KCO1. Plant J 29:809–820

12. Czempinski K, Gaedeke N, Zimmermann S, MuellerRoeber B (1999) Molecular mechanisms and regulation of plant ion channels. J Exp Bot 50:955–966

13. De Angeli A, Monachello D, Ephritikhine G, Frachisse JM, Thomine S, Gambale F, Barbier-Brygoo H (2006) The nitrate/proton antiporter AtCLCa mediates nitrate accumulation in plant vacuoles. Nature 442:939–942

14. De Angeli A, Moran O, Wege S, Filleur S, Ephritikhine G, Thomine S, BarbierBrygoo H, Gambale F (2009) ATP binding to the C terminus of the *Arabidopsis thaliana* nitrate/proton antiporter, AtCLCa, regulates nitrate transport into plant vacuoles. J Biol Chem 284:26526–26532

15. Doddema H, Telkamp GP (1979) Uptake of nitrate by mutants of *Arabidopsis thaliana* disturbed in uptake or reduction of nitrate. II. Kinetics. Physiol Plant 45:332–338

16. Forde BG (2000) Nitrate transporters in plants: structure, function and regulation. Biochem Biophys Acta 1465:219–235

17. Forde BG, Clarkson DT (1999) Nitrate and ammonium nutrition of plants: physiological and molecular perspectives. Adv Bot Res 30:1–90

18. Gaymard F, Pilot G, Lacombe B, Bouchez D, Bruneau D, Boucherez J, Michaux-Ferriere N, Thibaud JB, Sentenac H (1998) Identification and disruption of a plant shaker-like outward channel involved in K+ release into the xylem sap. Cell 94:647–655

19. Gierth M, Maser P (2007) Potassium transporters in plants—involvement in K^+ acquisition, redistribution and homeostasis. FEBS Lett 581:2348–2356

20. Gierth M, Maser P, Schroeder JI (2005) The potassium transporter AtHAK5 functions in K^+ deprivation induced high-affinity K^+ uptake and AKT1 K^+ channel contribution to K^+ uptake kinetics in Arabidopsis roots. Plant Physiol 137:1105–1114

21. Hashimoto K, Eckert C, Anschutz U, Scholz M, Held K, Waadt R, Reyer A, Hippler M, Becker D, Kudla J (2012) Phosphorylation of calcineurin B-like (CBL) calcium sensor proteins by their CBL-interacting protein kinases (CIPKs) is required for full activity of CBL–CIPK complexes toward their target proteins. J Biol Chem 287:7956–7968

22. Hedrich R, Kudla J (2006) Calcium signaling networks channel plant K^+ uptake. Cell 125:1221–1223

23. Held K, Pascaud F, Eckert C, Gajdanowicz P, Hashimoto K, Corratge-Faillie C, Offenborn JN, Lacombe B, Dreyer I, Thibaud JB, Kudla J (2011) Calcium-dependent modulation and plasma membrane targeting of the AKT2 potassium channel by the CBL4/CIPK6 calcium sensor/protein kinase complex. Cell Res 21:1116–1130

24. Ho CH, Lin SH, Hu HC, Tsay YF (2009) CHL1 functions as a nitrate sensor in plants. Cell 138:1184–1194
25. Hole DJ, Emran AM, Fares Y, Drew MC (1990) Induction of nitrate transport in maize roots, and kinetics of influx, measured with nitrogen-13. Plant Physiol 93:642–647
26. Hosy E, Vavasseur A, Mouline K, Dreyer I, Gaymard F, Poree F, Boucherez J, Lebaudy A, Bouchez D, Very AA et al. (2003) The Arabidopsis outward K+ channel GORK is involved in regulation of stomatal movements and plant transpiration. Proc Natl Acad Sci USA 100:5549–5554
27. Hu HC, Wang YY, Tsay YF (2009) AtCIPK8, a CBL-interacting protein kinase, regulates the low-affinity phase of the primary nitrate response. Plant J 57(2):264–278
28. Kim EJ, Kwak JM, Uozumi N, Schroeder JI (1998) AtKUP1: an Arabidopsis gene encoding high-affinity potassium transport activity. Plant Cell 10:51–62
29. Lan WZ, Lee SC, Che YF, Jiang YQ, Luan S (2011) Mechanistic analysis of AKT1 regulation by the CBL–CIPK–PP2CA interactions. Mol Plant 4:527–536
30. Lebaudy A, Very AA, Sentenac H (2007) K+ channel activity in plants: genes, regulations and functions. FEBS Lett 581:2357–2366
31. Lee SC, Lan WZ, Kim BG, Li L, Cheong YH, Pandey GK, Lu G, Buchanan BB, Luan S (2007) A protein phosphorylation/dephosphorylation network regulates a plant potassium channel. Proc Natl Acad Sci U S A 104:15959–15964
32. Leigh RA, Jones RGW (1984) A hypothesis relating critical potassium concentrations for growth to the distribution and functions of this ion in the plant-cell. New Phytol 97:1–13
33. Li L, Kim BG, Cheong YH, Pandey GK, Luan S (2006) A Ca^{2+} signaling pathway regulates a K^+ channel for low-K response in Arabidopsis. Proc Natl Acad Sci USA 103:12625–12630
34. Li W, Wang Y, Okamoto M, Nigel M, Crawford, Siddiqui MY, Glass ADM (2007) Dissection of the AtNRT2.1:AtNRT2.2 inducible high-affinity nitrate transporter gene cluster. Plant Physiol 143:425–433
35. Liu KH, Huang CY, Tsay YF (1999) CHL1 is a dual-affinity nitrate transporter of Arabidopsis involved in multiple phases of nitrate uptake. Plant Cell 11:865874
36. Liu LL, Ren HM, Chen LQ, Wang Y, Wu WH (2013) A protein kinase, calcineurin B-like protein-interacting protein Kinase9, interacts with calcium sensor calcineurin B-like Protein3 and regulates potassium homeostasis under low-potassium stress in Arabidopsis. Plant Physiol 161:266–277
37. Luan S (2009) The CBL–CIPK network in plant calcium signaling. Trends Plant Sci 14:37–42
38. Luan S, Lan W, Chul Lee S (2009) Potassium nutrition, sodium toxicity, and calcium signaling: connections through the CBL–CIPK network. Curr Opin Plant Biol 12:339–346
39. Marschner H (1995) Functions of mineral nutrients: macronutirents. In: Marschner H, Rimmington GM (eds) Mineral nutrition of higher plants, 2nd edn. Academic Press, New York, pp 299–312
40. Miller AJ, Fan X, Orsel M, Smith SJ, Wells DM (2007) Nitrate transport and signalling. J Exp Bot 58:2297–2306
41. Okamoto M, Kumar A, Li W, Wang Y, Siddiqi MY, Crawford NM, Glass AD (2006) High-affinity nitrate transport in roots of Arabidopsis depends on expression of the NAR2-like gene AtNRT3.11. Plant Physiol 140:1036–1046
42. Orsel M, Chopin F, Leleu O, Smith SJ, Krapp A, Daniel-Vedele F, Miller AJ (2006) Characterization of a two-component high-affinity nitrate uptake system in Arabidopsis. Physiology and protein–protein interaction. Plant Physiol 142:1304–1317
43. Pandey GK (2008) Emergence of a novel calcium signaling pathway in plants: CBL–CIPK signaling network. Physiol Mol Biol Plants 14:51–68
44. Pandey GK, Cheong YH, Kim BG, Grant JJ, Li L, Luan S (2007) CIPK9: a calcium sensor-interacting protein kinase required for low-potassium tolerance in Arabidopsis. Cell Res 17:411–421
45. Plett D, Toubia J, Garnett T, Tester M, Kaiser BN, Baumann U (2010) Dichotomy in the NRT gene families of dicots and grass species. PLoS ONE 5:e15289

46. Rubio F, Nieves-Cordones M, Aleman F, Martinez V (2008) Relative contribution of AtHAK5 and AtAKT1 to K^+ uptake in the high-affinity range of concentrations. Physiol Plant 134:598–608

47. Santa-Maria GE, Rubio F, Dubcovsky J, Rodriguez-Navarro A (1997) The HAK1 gene of barley is a member of a large gene family and encodes a high-affinity potassium transporter. Plant Cell 9:2281–2289

48. Schachtman DP (2000) Molecular insights into the structure and function of plant K^+ transport mechanisms. Biochim Biophys Acta 1465:127–139

49. Schachtman DP, Shin R (2007) Nutrient sensing and signaling: NPKS. Annu Rev Plant Biol 58:47–69

50. Siddiqi MY, Glass ADM, Ruth TJ, Fernando M (1989) Studies of the regulation of nitrate influx by barley seedlings using $^{13}NO_3^-$. Plant Physiol 90:806–813

51. Siddiqi MY, Glass ADM, Ruth TJ, Rufty TW (1990) Studies of the uptake of nitrate in barley. I. Kinetics of $^{13}NO^-$ influx. Plant Physiol 93:1426–1432

52. Tokas I, Pandey A, Pandey GK (2013) Role of calcium-mediated CBL–CIPK network in plant mineral nutrition and abiotic stress. Molecular stress physiology of plants, pp 241–261

53. Véry AA, Sentenac H (2003) Molecular mechanisms and regulation of K^+ transport in higher plants. Annu Rev Plant Biol 54:575–603

54. Von Wirén N, Gazzarrini S, Frommer WB (1997) Regulation of mineral nitrogen uptake in plants. Plant Soil 196:191–199

55. Walker DJ, Leigh RA, Miller AJ (1996) Potassium homeostasis in vacuolated plant cells. Proc Natl Acad Sci USA 93:10510–10514

56. Wang R, Crawford NM (1996) Genetic identification of a gene involved in constitutive, high-affinity nitrate transport in higher plants. Proc Natl Acad Sci USA 93:9297–9301

57. Wang R et al (1998) The Arabidopsis CHL1 protein plays a major role in high-affinity nitrate uptake. Proc Natl Acad Sci USA 95:15134–15139

58. Xu J, Li HD, Chen LQ, Wang Y, Liu LL, He L, Wu WH (2006) A protein kinase, interacting with two calcineurin B-like proteins, regulates K^+ transporter AKT1 in Arabidopsis. Cell 125:1347–1360

59. Yu Q, An L, Li W (2014) The CBL–CIPK network mediates different signaling pathways in plants. Plant Cell Rep 33:203–214

60. Zhang H, Lv F, Han X, Xia X, Yin W (2013) The calcium sensor PeCBL1, interacting with PeCIPK24/25 and PeCIPK26, regulates Na^+/K^+ homeostasis in Populus euphratica. Plant Cell Rep 32:611–621

61. Zhang H, Yin W, Xia X (2010) Shaker-like potassium channels in *Populus*, regulated by the CBL–CIPK signal transduction pathway, increase tolerance to low-K^+ stress. Plant Cell Rep 29:1007–1012

Chapter 8
Functional Role of CBL–CIPK in Abiotic Stresses

Abstract There are many environmental factors, directly or indirectly affecting the growth and development of the plants. Abiotic stresses are also designated as adverse environmental factors negatively impacting the growth and development and hence productivity of the crop plants. Because of rigorous anthropogenic development, more and more crops are venerable to abiotic stresses in the several parts of the world. The intrinsic property of plant to sense and respond to these adverse environmental cues enables them to protect through the activation of a large number of genes and gene networks. Signaling pathways are the crucial and key part of this intricate machinery involved in enabling plants to adapt and adjust in response to these abiotic stresses. Till date, calcium-mediated CBL–CIPK components are extensively studied to be involved in regulating several abiotic stress-triggered signaling cascades.

Keywords Function · Abiotic stress · Salt · Drought · Osmotic · Cold · ABA · Signaling · pH · Flooding · CBL · CIPK

8.1 Introduction

Among all the stresses to plant growth and development, abiotic stresses represent a major problem for agricultural practices worldwide. In response to salinity and dehydration or limited water availability, plant synthesizes ABA and triggers several cascades of signaling network. Elevation of cytosolic calcium concentration is a primary event in response to many different abiotic stresses such as salinity, osmotic stress, cold, dehydration, and water submergence (hypoxia). Although the specific signature of these calcium transients can encode information and specificity on its own, an additional level of regulation in calcium signaling is achieved via the action of calcium-binding proteins. In plants, three major calcium sensors regulate diverse arrays of signaling pathways that include calmodulin (CaM), calcineurin B-like (CBLs), and calcium-dependent protein kinases (CDPK). The role

© The Author(s) 2014   65
G.K. Pandey et al., *Global Comparative Analysis of CBL–CIPK Gene Families in Plants*, SpringerBriefs in Plant Science, DOI 10.1007/978-3-319-09078-8_8

of CBL–CIPK signaling regulating a diverse aspect of abiotic stresses has been adequately reported in Arabidopsis. In this chapter, we will make an attempt to elaborate several of these cases with examples.

8.2 Salt Stress

Salinity stress is caused by the presence of high concentration of salt in the soil, which imparts both osmotic and ionic stresses. Plant responds to salt stress by restricting the accumulation of toxic Na^+ ions to maintain the ion homeostasis [12, 52]. SOS pathway is a well-elucidated pathway in Arabidopsis for salt tolerance [28]. SOS signaling pathway has long been recognized as a key mechanism for Na^+ exclusion and ion homeostasis [64]. SOS pathway has three known components, the first calcium sensor, SOS3/CBL4 [27, 47, 66], its interacting kinase, SOS2/CIPK24 [19, 27, 47], and SOS1 a Na^+/H^+ antiporter, phosphorylated and activated by SOS3–SOS2 complex at plasma membrane [27, 44, 45, 48]. SOS pathway plays very important role in transducing Ca^{2+} signal, thereby maintaining ion homeostasis during salt stress and ultimately providing salt tolerance (Fig. 8.1) [37, 47, 48, 66].

Besides the SOS pathway, an alternative signaling pathway, i.e., CBL10–CIPK24/SOS2 complex that regulates a yet unknown vacuolar transporter/channel, to sequester Na^+ into the vacuolar compartments, is speculated to be responsible for Na^+ ion sequestration and detoxification in the plant cell (Fig. 8.1) [21, 30]. This fact suggests complex nature of salt stress signaling pathways regulated by multiple CBL–CIPK in plants.

Functionally, SOS pathway has been conserved in other plant species such as rice, apple, poplar, and tomato (Fig. 8.2). In rice, SOS pathway components consist of OsSOS3/OsCBL4, OsSOS2/OsCIPK24, and OsSOS1, which could complement yeast salt-sensitive mutants AXT3K when expressed together [40]. Moreover, AtSOS2/CIPK24 could phosphorylate OsSOS1 in an in vitro phosphorylation assay [40].

In poplar, putative SOS genes, i.e., *PtSOS1*, *PtSOS2*, and *PtSOS3*, responsive to salinity stress were reported to be involved in salt tolerance [51]. The localization of PtSOS1 and PtSOS3 was found to be similar to Arabidopsis SOS1 and SOS3 in the plasma membrane, and PtSOS2 was distributed throughout the cell [51]. Reconstitution of poplar PtSOS2 in *sos3* mutant leads to partial complementation, whereas there was no complementation seen in the case of *sos1* mutant, suggesting PtSOS2 functions downstream of SOS3 and upstream of SOS1. Interestingly, PtSOS3 interacted with and recruited PtSOS2 to the plasma membrane in yeast and also in plants, suggesting a high degree of functional conservation between Arabidopsis and poplar *SOS* genes.

The ectopic expression of *MdSOS2*, a *CIPK* gene from apple (*Malus domestica*), complemented the function of Arabidopsis *sos2* mutant and conferred enhanced salt tolerance to the transgenic Arabidopsis plants [24]. MdSOS2 is highly similar to

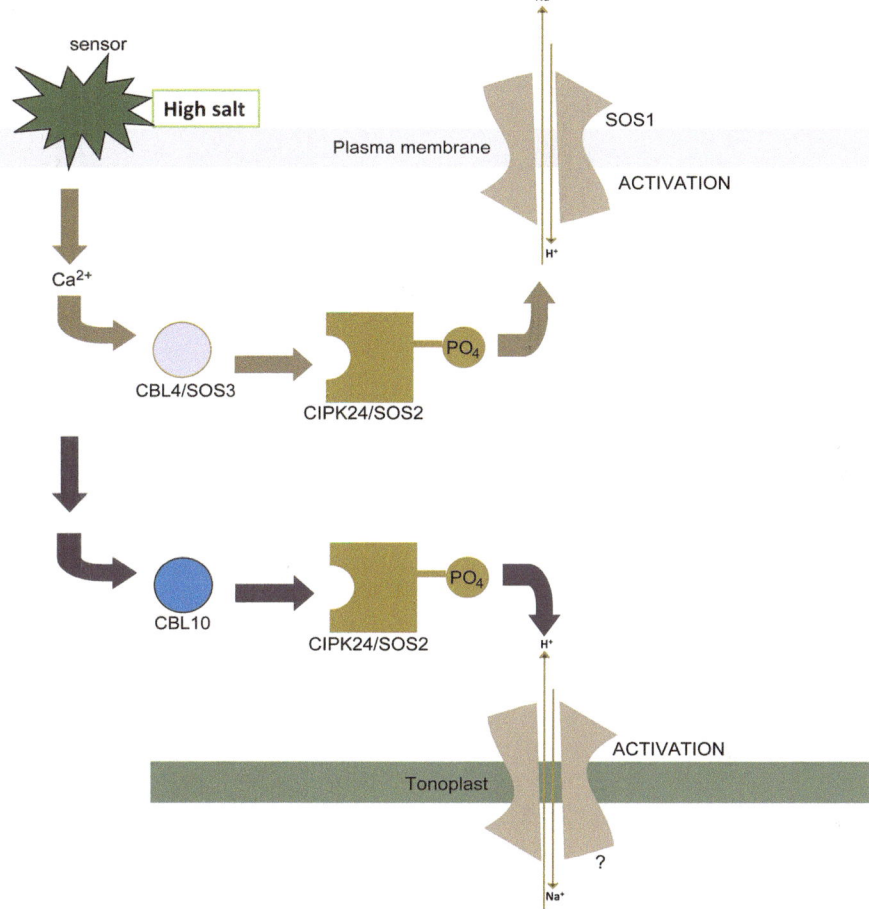

Fig. 8.1 Regulation of salt tolerance by CBL–CIPK signaling pathway. Two major pathways have been discovered, i.e., CBL4/SOS3–CIPK24/SOS2–Na^+/H^+ antiporter/SOS1 for efflux of salt through plasma membrane and CBL10–CIPK24–Unknown transporters at tonoplast for sequestering Na^+ into the vacuole

Arabidopsis SOS2 and N-terminal of MdSOS2 protein physically interacted with MdSOS3 and AtSOS3, respectively. In the case of tomatoes, there is no direct report to show that the SOS pathway functions in coordination [4, 26, 29]. Protein level of SlSOS3 in leaf increased when tomato plants were subjected to salt stress [4, 26, 29] and SlSOS2 overexpression conferred salt tolerance to transgenic tomato [26]. The increased salinity tolerance of *SlSOS2*-overexpressing plants was associated with higher sodium content in stems and leaves and with the induction and up-regulation of the SlSOS1 [4, 26, 29]. The plasma membrane Na^+/H^+ antiporter SOS1 is not only essential in maintaining ion homeostasis under salinity, but also critical for the partitioning of Na^+ between plant organs [26].

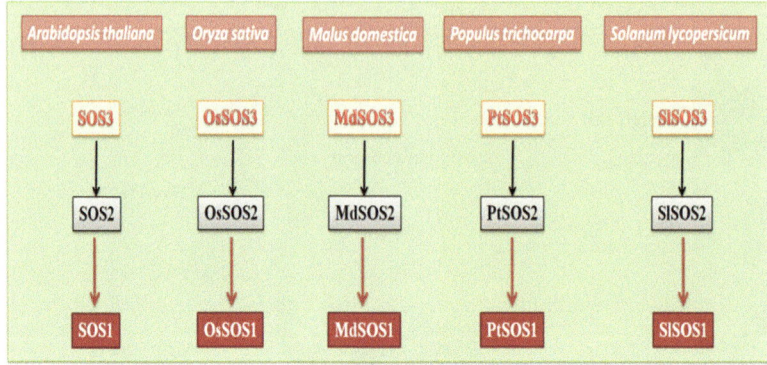

Fig. 8.2 Diagrammatic representation of SOS signaling pathway in different genera of plants. A high degree of functional conservation has been observed in components of salt tolerance pathway mediated by CBL–CIPK signaling components

Beside typical SOS components, i.e., CBL4 and CIPK24, there are other CBLs and CIPKs, which are also involved in salt stress signaling and tolerance [9, 11, 13, 46, 54]. Mutations in *CBL1* and *CBL9* lead to a hypersensitive response to grow on salt-containing media [9, 41], and overexpression of *CBL1* leads to enhanced salt tolerance [9]. Upon overexpression of *CBL5*, Arabidopsis plants are more tolerant to salt and osmotic stresses during seed germination [11]. In the case of CIPKs, loss of function of CIPK1 and CIPK6 led to sensitivity to growth on salt-containing media [13, 54]. In addition, overexpression of *CIPK6* and *CIPK16* in Arabidopsis leads to salt tolerance [7, 46]. Very recently, our group has identified another CIPK member; CIPK21 involved in the regulation of salt and osmotic stress responses by being preferentially targeted to vacuolar membrane (G.K. Pandey et al. unpublished data). The null mutation of CIPK21 leads to hypersensitivity response to grow on salt- and osmotic stress-containing media. At the same time, CIPK21 interacts specifically with CBL2 and CBL3 and targeted preferentially to tonoplast upon salinity condition (G.K. Pandey et al. unpublished data).

In rice, also there are other *CBLs* and *CIPKs*, and overexpression of a few of such candidates such as *OsCBL8* and *OsCIPK15* resulted in enhanced salt tolerance [55]. Similarly, in wheat, overexpression of *TaCIPK29* in tobacco resulted in increased salt tolerance [14]. Besides CBL4/SOS3–CIPK23/SOS2–SOS1 pathway, the complete elucidation and mechanism of salt tolerance have not been dissected for any other CBLs and CIPKs and still require an extensive amount of work.

8.3 Drought and Osmotic Stresses

Change in water potential is common effect of the high salinity and drought, which significantly leads to osmotic stress in the plants [2, 5, 58, 59]. Under osmotic stress, plant undergoes a number of physiological, developmental, and morphological changes in order to maintain the homeostasis and to detoxify harmful elements [32, 65]. As determined in earlier reports, abiotic stresses such as high salinity and drought are clearly different from each other in their physical nature and each elicits specific plant response; at the same time, they also activate some common responses in plants such as induction of common plant genes [9, 31, 66].

In Arabidopsis, role of CBL1, CBL5, and CBL9, and CIPK1, CIPK3, CIPK6, and CIPK16 has been implicated in drought and osmotic stress responses [1, 13, 41, 54]. CIPK23 has been shown to regulate the transpirational pull by regulating the opening and closing of stomata in ABA-dependent manner [10]. The null mutant alleles of *CIPK23* gene showed drought-tolerant phenotype. Moreover, CIPK23 was targeted to plasma membrane by CBL1 and CBL9, and both of these calcium sensors regulate CIPK23 synergistically since the double mutant of these calcium sensors *cbl1cbl9* but not the single mutants *cbl1* and *cbl9* showed drought-tolerant phenotype [10]. However, authors could not identify any downstream target of CIPK23 in the regulation of transpiration pull by opening and closing of stomata, and hence, the mechanism mediated by calcium CBL1/9–CIPK23 could not be completely addressed [10]. In the case of rice, OsCIPK23 has been shown to provide drought tolerance [62]. RNAi-mediated suppression of *OsCIPK23* expression conferred hypersensitivity to drought and expression of drought-tolerant genes in transgenic plants [62]. HbCIPK2, a novel CBL-interacting protein kinase from halophyte *Hordeum brevisubulatum*, confers salt and osmotic stress tolerance [36].

8.4 Cold Stress

A very few of the CBLs and CIPKs are implicated in cold stress responses by genetic and physiological analyses. One of the important candidates is *CBL1*; the null mutation of *CBL1* leads to enhanced cold and freezing tolerance, but opposite phenotype, i.e., sensitivity, has been noted for *CBL1*-overexpressing plants [9]. Moreover, the hyperinduction versus reduced induction of stress marker genes was observed in *CBL1* null mutant and overexpressing plants, respectively [1, 9]. Based on these reports, CBL1 was found to positively regulate salt and drought responses but negatively regulate cold and freezing responses [9]. In another report, CIPK7 was found to be associated with CBL1 in cold stress responses in Arabidopsis [25]. In rice (*Oryza sativa*), one of the *CIPKs*, i.e., *OsCIPK3*, was found to be highly up-regulated by cold stress and overexpression of *OsCIPK3* resulted in improved cold tolerance [55].

8.5 ABA Signaling

The role of ABA has been extensively studied in diverse growth and physiological pathways, especially during abiotic stress responses [15, 16, 23]. ABA acts as critical messenger for stress responses. Majorly, there are two types of ABA signaling pathway: ABA-dependent and ABA-independent regulatory pathways [16, 49, 56]. During several abiotic stress conditions, the biosynthesis of ABA is up-regulated to impart stress tolerance [57].

Pandey et al. [41] have shown that CBL9 negatively regulates ABA signaling and biosynthesis in Arabidopsis. The function of CBL9 was ascertained by reverse genetic approach where *CBL9* gene function was disrupted in Arabidopsis plants and the responses to ABA were drastically altered [41]. The mutant plants became hypersensitive to ABA in the early developmental stages, including during seed germination and post-germination seedling growth. In addition, seed germination in the mutant also showed increased sensitivity to inhibition by osmotic stress conditions produced by high concentrations of salt and mannitol. Further analyses indicated that increased stress sensitivity in the mutant might be because of both ABA hypersensitivity and increased accumulation of ABA under the stress conditions. Also, the expression of ABA and drought-responsive genes were hyper-induced in *cbl9* mutant. The *cbl9* mutant plants showed enhanced expression of genes involved in ABA signaling, such as ABA-INSENSITIVE 4 and 5.

Many members of CBL and CIPK families were characterized because of significant phenotypes observed for the loss-of-function mutants during different phenotypic screening experiments such as abiotic stress and ABA treatment in germination- or post-germination-based assays on MS plate or adult-stage stress tolerance assays. During ABA and abiotic stress treatment, one of the CIPKs family member, *CIPK3*, was found to be highly inducible at transcript level [31]. CIPK3, a Ser/Thr protein kinase that associates with a calcineurin B-like calcium sensor, regulates ABA response during seed germination and ABA- and stress-induced gene expression in Arabidopsis. *CIPK3* altered the expression pattern of a number of stress gene markers in response to ABA, cold, and high salt. However, drought-induced gene expression was not altered in the *cipk3* mutant plants, suggesting that *CIPK3* regulates select pathways in response to abiotic stress and ABA. These results identify *CIPK3* as a molecular link between stress- and ABA-induced calcium signal and gene expression in plant cells. Because the cold signaling pathway is largely independent of endogenous ABA production, CIPK3 represents a cross talk 'node' between the ABA-dependent and ABA-independent pathways in stress responses. Single mutant analysis of *CBL9* [41] and *CIPK3* [31] implicated the role of these proteins in ABA and osmotic stress signaling pathways. Further, through genetic and interaction studies, it was found that CBL9-CIPK3 pair function in same pathway that negatively regulates ABA response during seed germination in Arabidopsis (Pandey et al. 2008).

In a parallel study by Guo et al. [22], the function of SCaBP5/CBL1–PKS3/CIPK15 pair in ABA signaling pathway has also been determined. RNA silencing of both *SCaBP5* and *PKS3* leads to ABA hypersensitive phenotype in the silenced transgenic lines. In the biochemical analysis of SCaBP5 and PKS3, the interaction between these proteins was determined. Double mutant analysis of these did not produce cumulative phenotype, further confirming that they act in the same pathway. In a step further, ABI1/2 was found to interact physically with PKS3/CIPK15 and the dominant *abi1-1* and *abi2-1* mutant could suppress the *scabp5* and *pks3* hypersensitive response to seed germination and seedling growth [22]. Although mechanistic detail has not been found but based on the interaction between PKS3 and ABI1/2, and suppression of hypersensitive phenotype of *scabp5* and *pks3* mutants, they suggested that ScaBP5, PKS3, and ABI1/2 all work in the same pathway. Because CBL1/SCaBP5 and CBL9 share very high level of homology (almost 90 % identical at amino acid level), the use of coding region for RNA interference might have silenced both CBL1/SCaBP5 and CBL9 and hence leads to this hypersensitive phenotype. On the basis of single knockout mutant analysis of *CBL1* [1, 9] and *CBL9* [41], it was found that CBL1 function in regulating dehydration, osmotic stress, whereas, CBL9 in regulating ABA specific pathway [9, 41]. The complexity of CBL1 and CBL9 functions is further investigated by creating *cbl1cbl9* double mutant, which showed synergistic regulation of ABA-mediated stomatal responses and low-K sensitivity/uptake by interacting with CIPK23 and drought tolerance phenotype [10]. The role of an AP2 domain/ERBP-containing transcription factor, AtERF7, was investigated where it interacts physically with PKS3/CIPK15 [50]. PKS3/CIPK15 phosphorylates AtERF7 in vitro and the genetic analysis of RNA-silenced lines also implicated the mutant of *AtERF7* in ABA hypersensitive responses in seed germination and seedling growth [50]. AtERF7 along with two other proteins, AtSIN3 and HDA19, was implicated as repressors of ABA-mediated gene expression [50].

Similarly, in a microarray analysis of *cipk3*, several genes were found consistently up- or down-regulated in *cipk3* mutant (J.J. Grant, G.K. Pandey, S. Luan, unpublished). One of the important genes identified as target of CIPK3 was AP2 domain-containing transcription factor, which was down-regulated more than twofold in *cipk3* mutant under ABA treatment (J.J. Grant, G.K. Pandey, and S. Luan, unpublished). Detailed reverse genetic analysis of this AP2 domain-containing mutant, *abr1*, showed similar ABA-sensitive phenotype as *cipk3* mutant [42]. The recent identification of direct interaction between CIPK3 and ABR1, and phosphorylation of ABR1 by CIPK3 suggests the possible regulation of ABA-mediated signaling by calcium CBL9–CIPK3–ABR1 signaling network (S. Sanyal and G.K. Pandey, Unpublished data). In a study by Gong et al. [20], *CIPK20* was found to regulate ABA responses where the RNAi-silencing lines showed insensitivity to ABA. However, the overexpression of CIPK20 in Arabidopsis resulted in hypersensitivity response to ABA during seed germination and seedling growth [20].

Recently, CIPK26 was found to regulate ABA signaling pathway positively in Arabidopsis [39]. Overexpression of *CIPK26* induces an ABA hypersensitive phenotype, and it interacts with RING-type E3 ligase Keep on Going (KEG), a negative regulator of ABA signaling [39]. CIPK26 interacts with ABI5, confirmed by biochemical evidence. Interestingly, KEG protein targets ABI5 and CIPK26 for degradation, while CIPK26 was shown to phosphorylate and regulate the activity of ABI5, suggesting that KEG is acting to counteract and balance ABA responses, which are promoted by CIPK26 and ABI5 (Fig. 8.3).

Mutant analyses of *cbl2* knockout revealed that CBL2 function is required for appropriate ABA responses during seed germination [3]. The S-acylation of

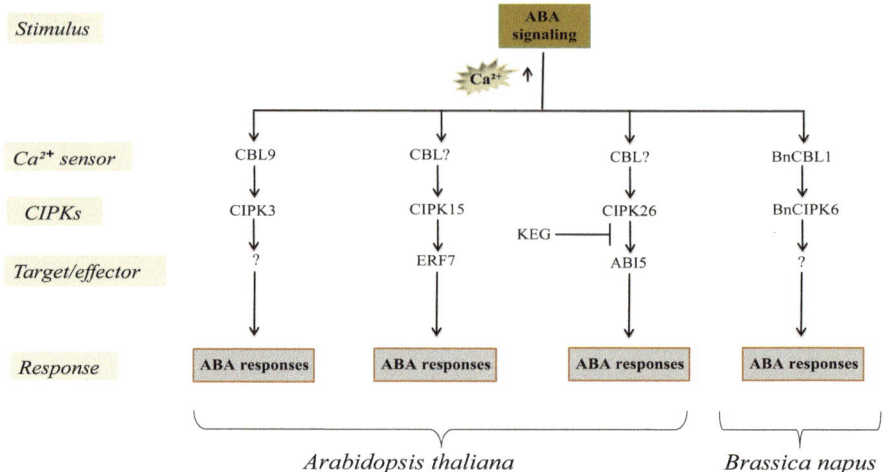

Fig. 8.3 Involvement of CBL–CIPK signaling pathways in ABA signaling. The diagram represents the four ABA signaling pathways regulated by different CBL–CIPK modules in Arabidopsis

CBL2 and proper tonoplast targeting are required for full functionality of CBL2 as tested by the complementation of *cbl2* mutant. However, the regulation mechanism of ABA responses at the vacuole by CBL2 is still unknown [3]. In *Brassica napus,* CBL1 and CIPK6 are involved in the plant responses to abiotic stresses and ABA signaling [6].

8.6 pH Stress

Alkaline or acidic soil also imparts abiotic stress and causes severe problem for agricultural productivity. The regulatory pathways responsible for pH stress responses are not yet understood. Since pH is very important parameter for cellular activities in the plant cell, the drastic imbalance in external pH can affect the growth and development of plants. In the calcium-mediated signaling pathways, CBL2 recruits CIPK11/PKS5 to the plasma membrane to phosphorylate and negatively regulate PM H^+-ATPase and AHA2 and regulate the cytoplasmic pH homeostasis [17, 63]. The loss-of-function of *CIPK11* in Arabidopsis resulted in plant tolerance to external higher pH [17]. Phosphorylation by CBL2–CIPK11/PKS5 complex at C-terminal of AHA2 at Ser931 prevents interaction of activating protein 14-3-3 with PM H^+-ATPase [17].

According to another report by Yang et al. [63], a chaperone, J3 (DnaJ homolog 3), physically interacts with CIPK11/PKS5. *j3* mutant plants result in hypersensitivity to salt at high external pH. Not only this, but also PM H^+-ATPase activity was decreased in the mutant. The double mutant *j3–1pks5* showed similar responses to salt at alkaline pH like the corresponding *j3* and *pks5* single mutants, suggesting that they function in the same genetic pathway. From this study, it was concluded that J3 functions upstream of PKS5/CIPK11 and regulates PM H$^+$-ATPase activity via inactivation of the PKS5/CIPK11 kinase [63].

CIPK11/PKS5 and J3 regulate the interaction between 14-3-3 proteins and PM ATPase [17, 63]. Therefore, Xu et al. in [61] studied the role of the 14-3-3 genes in *Solanum lycopersicum*. The gene expression of four tomatoes 14-3-3 proteins (*TFT1*, *TFT4*, *TFT6*, and *TFT7*) was enhanced in alkaline stress. Out of these four genes, tolerance to alkaline stress was found in the *TFT4*-overexpressing Arabidopsis plants. The role of TFT4 in tomato has been shown to regulate the activity of PM H$^+$-ATPase in alkaline stress and IAA transport [61]. Overall, TFT4 acts as a regulator of the PKS5/CIPK11–J3 pathway in tomato and alkaline stress responses in the root apex [61].

8.7 Flooding Stress

Water limitation results in drought stress; similarly, excess water leads to submergence or flooding stress, creating oxygen deprivation conditions. Flooding stress alters the respiration physiology and metabolic pathways. Like many other food crops, rice (*Oryza sativa* L.) is also susceptible to flooding stress, and most of the arable coastal areas in the world are affected by flooding stress. Not many genes and gene products have been functionally characterized to regulate flooding stress in crop plants, especially in the context of providing flooding tolerance. One of the major and important genes identified to provide a dramatic flooding tolerance is Sub1A also known as Submergence 1A. *Sub1A* encodes for ethylene-responsive factor (ERF) gene in rice indica varieties and is believed to be responsible for the submergence tolerance [60]. Sub1A negatively regulates carbohydrate catabolism, including α-amylase-dependent starch degradation [18]. Submergence 1A, induction during flooding, results in quiescence status because of inhibition of carbohydrate catabolism and elongation of the submerged plants. And this subsequently leads to faster recovery during the onset of de-submergence in plants.

Ramy3D is an α-amylase, which plays a major role for low oxygen survival since it is highly induced under low oxygen [34, 43]. CIPK15 and SNF1-related protein kinase 1A (SnRK1A) activate MYBS1 that up-regulates *Ramy3D* [8, 35, 38, 53]. CIPK15-mediated regulation of α-amylase pathway is repressed in the tolerant, Sub1A-containing FR13A variety of rice under O$_2$ deprivation [33, 60]. It is still unknown whether Sub1A- and CIPK15-mediated pathways act as complementary processes for rice survival under O$_2$ deprivation. Overall, Sub1A acts as a starch degradation repressor and CIPK15 as an activator. Hence, calcium signaling either through CBL–CIPK or through directly regulating the Sub1A in the carbohydrate consumption under oxygen deprivation might be associated with the regulation of flooding stress responses.

References

1. Albrecht V, Weinl S, Blazevic D, D'Angelo C, Batistic O, Kolukisaoglu U, Bock R, Schulz B, Harter K, Kudla J (2003) The calcium sensor CBL1 integrates plant responses to abiotic stresses. Plant J 36:457–470
2. Bartels D, Sunkar R (2005) Drought and salt tolerance in plants. Crit Rev Plant Sci 24:236–241
3. Batistic O, Rehers M, Akerman A, Schlucking K, Steinhorst L, Yalovsky S, Kudla J (2012) S-acylation-dependent association of the calcium sensor CBL2 with the vacuolar membrane is essential for proper abscisic acid responses. Cell Res 22:1155–1168
4. Belver A, Olias R, Huertas R, Rodriguez-Rosales MP (2012) Involvement of SlSOS2 in tomato salt tolerance. Bioengineered 3:298–302
5. Boudsocq M, Lauriere C (2005) Osmotic signaling in plants: multiple pathways mediated by emerging kinase families. Plant Physiol 138:1185–1194
6. Chen PW, Lu CA, Yu TS, Tseng TH, Wang CS, Yu SM (2002) Rice α-amylase transcriptional enhancers direct multiple mode regulation of promoters in transgenic rice. J Biol Chem 277:13641–13649
7. Chen L, Ren F, Zhou L, Wang QQ, Zhong H, Li XB (2012) The *Brassica napus* calcineurin B-Like 1/CBL-interacting protein kinase 6 (CBL1/CIPK6) component is involved in the plant response to abiotic stress and ABA signalling. J Exp Bot 63:6211–6222
8. Chen L, Wang QQ, Zhou L, Ren F, Li DD, Li XB (2013) Arabidopsis CBL-interacting protein kinase (CIPK6) is involved in plant response to salt/osmotic stress and ABA. Mol Biol Rep 40:4759–4767
9. Cheong YH, Kim KN, Pandey GK, Gupta R, Grant JJ, Luan S (2003) CBL1, a calcium sensor that differentially regulates salt, drought, and cold responses in Arabidopsis. Plant Cell 15:1833–1845
10. Cheong YH, Pandey GK, Grant JJ, Batistic O, Li L, Kim BG, Lee SC, Kudla J, Luan S (2007) Two calcineurin B-like calcium sensors, interacting with protein kinase CIPK23, regulate leaf transpiration and root potassium uptake in Arabidopsis. Plant J 52:223–239
11. Cheong YH, Sung SJ, Kim BG, Pandey GK, Cho JS, Kim KN, Luan S (2010) Constitutive overexpression of the calcium sensor CBL5 confers osmotic or drought stress tolerance in Arabidopsis. Mol Cells 29:159–165
12. Clarkson DT, Hanson JB (1980) The mineral nutrition of higher-plants. Ann Rev Plant Phys 31:239–298
13. Deng X, Hu W, Wei S, Zhou S, Zhang F, Han J, Chen L, Li Y, Feng J, Fang B, Luo Q, Li S, Liu Y, Yang G, He G (2013) TaCIPK29, a CBL-interacting protein kinase gene from wheat, confers salt stress tolerance in transgenic tobacco. PLoS ONE 8:e69881
14. D'Angelo C, Weinl S, Batistic O, Pandey GK, Cheong YH, Schultke S, Albrecht V, Ehlert B, Schulz B, Harter K, Luan S, Bock R, Kudla J (2006) Alternative complex formation of the Ca-regulated protein kinase CIPK1 controls abscisic acid-dependent and independent stress responses in Arabidopsis. Plant J 48:857–872
15. Fedoroff NV (2002) Cross-talk in abscisic acid signaling. Sci STKE RE10
16. Finkelstein RR, Gampala SSL, Rock CD (2002) Abscisic acid signaling in seeds and seedlings. Plant Cell 14:S15–S45
17. Fuglsang AT, Guo Y, Cuin TA, Qiu Q, Song C, Kristiansen KA, Bych K, Schulz A, Shabala S, Schumaker KS, Palmgren MG, Zhu JK (2007) Arabidopsis protein kinase PKS5 inhibits the plasma membrane H^+-ATPase by preventing interaction with 14-3-3 protein. Plant Cell. 19:1617–1634
18. Fukao T, Xu K, Ronald PC, Bailey-Serres J (2006) A variable cluster of ethylene response factor-like genes regulates metabolic and developmental acclimation responses to submergence in rice. Plant Cell 18:2021–2034
19. Gong D, Guo Y, Schumaker KS, Zhu JK (2004) The SOS3 family of calcium sensors and SOS2 family of protein kinases in Arabidopsis. Plant Physiol 134:919–926

20. Gong D, Zhang C, Chen X, Gong Z, Zhu JK (2002) Constitutive activation and transgenic evaluation of the function of an Arabidopsis PKS protein kinase. J Biol Chem 277:42088–42096
21. Guo Y, Halfter U, Ishitani M, Zhu JK (2001) Molecular characterization of functional domains in the protein kinase SOS2 that is required for plant salt tolerance. Plant Cell 13:1383–1400
22. Guo Y, Xiong L, Song CP, Gong D, Halfter U, Zhu JK (2002) A calcium sensor and its interacting protein kinase are global regulators of abscisic acid signaling in Arabidopsis. Dev Cell 3:233–244
23. Himmelbach A, Yang Y, Grill E (2003) Relay and control of abscisic acid signaling. Curr Opin Plant Biol 6:470–479
24. Hu DG, Li M, Luo H, Dong QL, Yao YX, You CX, Hao YJ (2012) Molecular cloning and functional characterization of MdSOS2 reveals its involvement in salt tolerance in apple callus and Arabidopsis. Plant Cell Rep 31:713–722
25. Huang C, Ding S, Zhang H, Du H, An L (2011) CIPK7 is involved in cold response by interacting with CBL1 in *Arabidopsis thaliana*. Plant Sci 181:57–64
26. Huertas R, Olias R, Eljakaoui Z, Galvez FJ, Li J, De Morales PA, Belver A, Rodriguez-Rosales MP (2012) Overexpression of SlSOS2 (SlCIPK24) confers salt tolerance to transgenic tomato. Plant Cell Environ 35:1467–1482
27. Ishitani M, Liu J, Halfter U, Kim CS, Shi W, Zhu JK (2000) SOS3 function in plant salt tolerance requires N-myristoylation and calcium binding. Plant Cell 12:1667–1678
28. Ji H, Pardo JM, Batelli G, Van Oosten MJ, Bressan RA, Li X (2013) The salt overly sensitive (SOS) pathway: established and emerging roles. Mol Plant 6:275–286
29. Kim KN, Cheong YH, Grant JJ, Pandey GK, Luan S (2003) CIPK3, a calcium sensor-associated protein kinase that regulates abscisic acid and cold signal transduction in Arabidopsis. Plant Cell 15:411–423
30. Kim BG, Waadt R, Cheong YH, Pandey GK, Dominguez-Solis JR, Schultke S, Lee SC, Kudla J, Luan S (2007) The calcium sensor CBL10 mediates salt tolerance by regulating ion homeostasis in Arabidopsis. Plant J 52:473–484
31. Krasensky J, Jonak C (2012) Drought, salt, and temperature stress-induced metabolic rearrangements and regulatory networks. J Exp Bot 63:1593–1608
32. Kudahettige NP, Pucciariello C, Parlanti S, Alpi A, Perata P (2011) Regulatory interplay of the Sub1A and CIPK15 pathways in the regulation of α-amylase production in flooded rice plants. Plant Biol 13:611–619
33. Lasanthi-Kudahettige R, Magneschi L, Loreti E, Gonzali S, Licausi F, Novi G, Beretta O, Vitulli F, Alpi A, Perata P (2007) Transcript profiling of the anoxic rice coleoptile. Plant Physiol 144:218–231
34. Lee KW, Chen PW, Lu CA, Chen S, Ho THD, Yu SM (2009) Coordinated responses to oxygen and sugar deficiency allow rice seedlings to tolerate flooding. Sci Signal 2:ra61
35. Li R, Zhang J, Wu G, Wang H, Chen Y, Wei J (2012) HbCIPK2, a novel CBL-interacting protein kinase from halophyte *Hordeum brevisubulatum*, confers salt and osmotic stress tolerance. Plant Cell Environ 35:1582–1600
36. Liu J, Ishitani M, Halfter U, Kim CS, Zhu JK (2000) The *Arabidopsis thaliana* SOS2 gene encodes a protein kinase that is required for salt tolerance. Proc Natl Acad Sci USA 97:3730–3734
37. Lu CA, Lim EK, Yu SM (1998) Sugar response sequence in the promoter of a rice α-amylase gene serves as a transcriptional enhancer. J Biol Chem 273:10120–10131
38. Lyzenga WJ, Liu H, Schofield A, Muise-Hennessey A, Stone SL (2013) Arabidopsis CIPK26 interacts with KEG, components of the ABA signalling network and is degraded by the ubiquitin-proteasome system. J Exp Bot 64:2779–2791
39. Martinez-Atienza J, Jiang X, Garciadeblas B, Mendoza I, Zhu JK, Pardo JM, Quintero FJ (2007) Conservation of the salt overly sensitive pathway in rice. Plant Physiol 143:1001–1012

40. Pandey GK, Cheong YH, Kim KN, Grant JJ, Li L, Hung W, D'Angelo C, Weinl S, Kudla J, Luan S (2004) The calcium sensor calcineurin B-like 9 modulates abscisic acid sensitivity and biosynthesis in Arabidopsis. Plant Cell 16:1912–1924

41. Pandey GK, Grant JJ, Cheong YH, Kim BG, Li L, Luan S (2005) ABR1, an APETALA2-domain transcription factor that functions as a repressor of ABA response in Arabidopsis. Plant Physiol 139:1185–1193

42. Pandey GK, Grant JJ, Cheong YH, Kim BG, le Li G, Luan S (2008) Calcineurin-B-like protein CBL9 interacts with target kinase CIPK3 in the regulation of ABA response in seed germination. Mol Plant 1:238--24

43. Perata P, Matsukura C, Vernieri P, Yamaguchi J (1997) Sugar repression of a gibberellin-dependent signaling pathway in barley embryos. Plant Cell 9:2197–2208

44. Qiu QS, Guo Y, Dietrich MA, Schumaker KS, Zhu JK (2002) Regulation of SOS1, a plasma membrane Na$^+$/H$^+$ exchanger in *Arabidopsis thaliana*, by SOS2 and SOS3. Proc Natl Acad Sci USA 99:8436–8441

45. Qiu QS, Guo Y, Quintero FJ, Pardo JM, Schumaker KS, Zhu JK (2004) Regulation of vacuolar Na$^+$/H$^+$ exchange in *Arabidopsis thaliana* by the salt-overly-sensitive (SOS) pathway. J Biol Chem 279:207–215

46. Roy SJ, Huang W, Wang XJ, Evrard A, Schmockel SM, Zafar ZU, Tester M (2013) A novel protein kinase involved in Na$^+$ exclusion revealed from positional cloning. Plant Cell Environ 36:553–568

47. Sanchez-Barrena MJ, Martinez-Ripoll M, Zhu JK, Albert A (2005) The structure of the *Arabidopsis thaliana* SOS3: molecular mechanism of sensing calcium for salt stress response. J Mol Biol 345:1253–1264

48. Shi H, Ishitani M, Kim C, Zhu JK (2000) The *Arabidopsis thaliana* salt tolerance gene SOS1 encodes a putative Na$^+$/H$^+$ antiporter. Proc Natl Acad Sci USA 97:6896–6901

49. Shinozaki K, Yamaguchi-Shinozaki K (2000) Molecular responses to dehydration and low temperature: differences and cross-talk between two stress signaling pathways. Curr Opin Plant Biol 3:217–223

50. Song CP, Agarwal M, Ohta M, Guo Y, Halfter U, Wang P, Zhu JK (2005) Role of an Arabidopsis AP2/EREBP-type transcriptional repressor in abscisic acid and drought stress responses. Plant Cell 17:2384–2396

51. Tang RJ, Liu H, Bao Y, Lv QD, Yang L, Zhang HX (2010) The woody plant poplar has a functionally conserved salt overly sensitive pathway in response to salinity stress. Plant Mol Biol 74:367–380

52. Tester M, Davenport R (2003) Na$^+$ tolerance and Na$^+$ transport in higher plants. Ann Bot 91:503–527

53. Toyofuku K, Umemura T, Yamaguchi J (1998) Promoter elements required for sugar-repression of theRAmy3D gene for α-amylase in rice. FEBS Lett 428:275–280

54. Tripathi V, Parasuraman B, Laxmi A, Chattopadhyay D (2009) CIPK6, a CBL-interacting protein kinase is required for development and salt tolerance in plants. Plant J 58:778–790

55. Xiang Y, Huang Y, Xiong L (2007) Characterization of stress-responsive CIPK genes in rice for stress tolerance improvement. Plant Physiol 144:1416–1428

56. Xiong L, Ishitani M, Zhu JK (1999) Interaction of osmotic stress, temperature, and abscisic acid in the regulation of gene expression in Arabidopsis. Plant Physiol 119:205–212

57. Xiong L, Schumaker KS, Zhu JK (2002) Cell signaling during cold, drought, and salt stress. Plant Cell 14:S165–S183

58. Xiong L, Zhu JK (2002) Molecular and genetic aspects of plant responses to osmotic stress. Plant Cell Environ 25:131–139

59. Xiong L, Zhu JK (2003) Regulation of abscisic acid biosynthesis. Plant Physiol 133:29–36

60. Xu W, Jia L, Shi W, Baluska F, Kronzucker HJ, Liang J, Zhang J (2013) The Tomato 14-3-3 protein TFT4 modulates H$^+$ efflux, basipetal auxin transport, and the PKS5-J3 pathway in the root growth response to alkaline stress. Plant Physiol 163:1817–1828

61. Xu K, Xu X, Fukao T, Canlas P, Maghirang-Rodriguez R, Heuer S, Ismail AM, Bailey-Serres J, Ronald PC, Mackill DJ (2006) Sub1A is an ethylene-response-factor-like gene that confers submergence tolerance to rice. Nature 442:705–708

62. Yang Y, Qin Y, Xie C, Zhao F, Zhao J, Liu D, Chen S, Fuglsang AT, Palmgren MG, Schumaker KS, Deng XW, Guo Y (2010) The Arabidopsis chaperone J3 regulates the plasma membrane H^+-ATPase through interaction with the PKS5 kinase. Plant Cell 22:1313–1332

63. Yang W, Kong Z, Omo-Ikerodah E, Xu W, Li Q, Xue Y (2008) Calcineurin B-like interacting protein kinase OsCIPK23 functions in pollination and drought stress responses in rice (*Oryza sativa* L.). J Genet Genomics 35:531–543, s531–s532

64. Zhu JK (2000) Genetic analysis of plant salt tolerance using Arabidopsis. Plant Physiol 124:941–948

65. Zhu JK (2001) Plant salt tolerance. Trends Plant Sci 6:66–71

66. Zhu JK (2002) Salt and drought stress signal transduction in plants. Annu Rev Plant Biol 53:247–273

67. Zhu JK, Liu J, Xiong L (1998) Genetic analysis of salt tolerance in Arabidopsis. Evidence for a critical role of potassium nutrition. Plant Cell 10:1181–1191

Chapter 9
Functional Role of CBL–CIPK in Biotic Stress and ROS Signaling

Abstract One important feature distinguishing plants from other complex multi-cellular organisms is that the plants are sessile and thus have to endure environmental challenges such as abiotic and biotic stresses. Although abiotic and biotic stresses are clearly different from each other in their nature and each elicits specific plant responses, they also activate some common reactions in plants. In addition, there are other second messengers such as calcium, reactive oxygen species (ROS), and phytohormone like jasmonic acid, which were shown to be implicated both in abiotic and biotic stresses. So detailed analysis of CBL–CIPK in biotic stress and their relationship with abiotic stress need to be explored in plants.

Keywords Function · Biotic stress · ROS · Signaling · CBL · CIPK

9.1 Introduction

Stress can be imposed by biological and non-biological components such as biotic and abiotic, respectively, which ultimately limit crop productivity worldwide. Understanding plant responses to these stresses is essential for rational engineering of crop plants. In Arabidopsis, the signal transduction pathways for abiotic stresses, light, several phytohormones, and pathogenesis are well understood. Moreover, there are several molecular 'Nodes and Hub' components, which are involved in many of these stresses. Plant responses to these stresses involve nearly every aspect of plant physiology and metabolism. Consequently, there exists a complex signaling network underlying plant adaptation to these adverse environmental and biotic factors. A significant portion of Arabidopsis and many other plant genomes encodes for genes involving in signaling components such as sensors, kinases, phosphatases, and transcription factors. The most widely studied common response is the induction of several plant genes by abiotic and biotic stresses. Involvement of CBL–CIPK in biotic stress is still not a much explored area compared to abiotic stresses, but some reports are coming up in this field too.

© The Author(s) 2014   79
G.K. Pandey et al., *Global Comparative Analysis of CBL–CIPK Gene Families in Plants*, SpringerBriefs in Plant Science, DOI 10.1007/978-3-319-09078-8_9

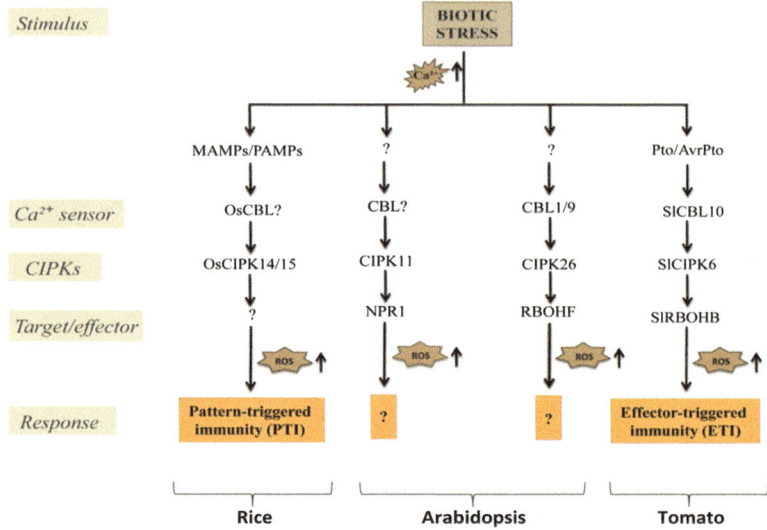

Fig. 9.1 Involvement of CBL–CIPK signaling pathways in biotic stress. The diagram represents the various CBL–CIPK-mediated signaling pathways present in different plant species regulating the biotic stress responses

9.2 Biotic Stress and ROS Signaling

In several plant–pathogen interaction and challenge responses, Ca^{2+} signaling is also shown to be involved in activating the defense responses. In many cases, elevation of $[Ca^{2+}]_{cyt}$ and synthesis of reactive oxygen species (ROS) are prerequisite for establishing pathogen resistance [5]. Respiratory burst oxidase homologs (RBOH) are known as ROS-producing enzymes [9, 10], and RBOHF has been implicated in response to biotic stresses [2, 13, 14]. Arabidopsis CIPK26 specifically interacts with the N-terminal domain of RBOHF in yeast two-hybrid analyses and with the full-length RBOHF protein in plant cells, and also phosphorylates RBOHF in in vitro conditions [4]. Co-expression of either *CBL1* or *CBL9* with *CIPK26* strongly enhances ROS production by RBOHF in HEK293T cells (Fig. 9.1) [4]. In another report, non-expressor of pathogenesis-related gene 1 (NPR1), which is a major co-activator of plant defense, interacts with PKS5/CIPK11 and phosphorylated at the C-terminal region (Fig. 9.1) [6]. Moreover, PKS5/CIPK11 functions upstream of NPR1 and might also mediate expression of *WRKY38* and *WRKY62* in disease response mechanism [11].

Pattern-triggered immunity (PTI) and effector-triggered immunity (ETI) are two very important part of the plant pathogen interaction. PTI results from the recognition of microbe- or pathogen-associated molecular patterns (MAMPs/PAMPs). PTI is accompanied by a localized programmed cell death (PCD) event known as the hypersensitive response (HR), which is believed to limit pathogen establishment and spread by killing both the pathogen and host cell [7], whereas ETI occurs when cytoplasmic resistance (R) proteins detect specific pathogen effectors.

In rice, *OsCIPK14* and *OsCIPK15* were found to be induced during MAMPs [8]. Silencing of *OsCIPK14/15* results in the suppression of the MAMP-induced ROS production as well as cell browning, which suggest the role of OsCIPK14/15 in regulating the PTI (Fig. 9.1) [8]. The tomato (*Solanum lycopersicum*) kinase Pto triggers localized PCD upon recognition of *Pseudomonas syringae* effectors AvrPto or AvrPtoB [1]. Overexpression of tomato *CIPK6* in *Nicotiana benthamiana* leaves causes accumulation of ROS, which requires the respiratory burst homolog RbohB [12]. Tomato CIPK6 interacts with CBL10, and silencing of these genes inhibits Pto/AvrPto-elicited PCD [1, 12]. Moreover, tomato CBL10–CIPK6 complex interacted with RbohB at the plasma membrane and contributed to ROS generation during effector-triggered immunity in the interaction of *P. syringae* pv tomato DC3000 (Fig. 9.1) [3]. In addition, one of the CIPK of wheat, i.e., TaCIPK29, was also found to be involved in ROS homeostasis in plant cell [3].

References

1. Abramovitch RB, Martin GB (2005) AvrPtoB: a bacterial type III effector that both elicits and suppresses programmed cell death associated with plant immunity. FEMS Microbiol Lett 245:1–8
2. Chaouch S, Queval G, Noctor G (2012) AtRbohF is a crucial modulator of defence-associated metabolism and a key actor in the interplay between intracellular oxidative stress and pathogenesis responses in Arabidopsis. Plant J 69:613–627
3. Deng X, Hu W, Wei S, Zhou S, Zhang F, Han J, Chen L, Li Y, Feng J, Fang B, Luo Q, Li S, Liu Y, Yang G, He G (2013) TaCIPK29, a CBL-interacting protein kinase gene from wheat, confers salt stress tolerance in transgenic tobacco. PLoS ONE 8:e69881
4. Drerup MM, Schlucking K, Hashimoto K, Manishankar P, Steinhorst L, Kuchitsu K, Kudla J (2013) The calcineurin B-like calcium sensors CBL1 and CBL9 together with their interacting protein kinase CIPK26 regulate the Arabidopsis NADPH oxidase RBOHF. Mol Plant 6:559–569
5. Dubiella U, Seybold H, Durian G, Komander E, Lassig R, Witte CP, Schulze WX, Romeis T (2013) Calcium-dependent protein kinase/NADPH oxidase activation circuit is required for rapid defense signal propagation. Proc Natl Acad Sci USA 110:8744–8749
6. Fuglsang AT, Guo Y, Cuin TA, Qiu Q, Song C, Kristiansen KA, Bych K, Schulz A, Shabala S, Schumaker KS, Palmgren MG, Zhu JK (2007) Arabidopsis protein kinase PKS5 inhibits the plasma membrane H^+-ATPase by preventing interaction with 14-3-3 protein. Plant Cell 19:1617–1634
7. Greenberg JT (1997) Programmed cell death in plant–pathogen interactions. Annu Rev Plant Physiol Plant Mol Biol 48:525–545
8. Kurusu T, Hamada J, Nokajima H, Kitagawa Y, Kiyoduka M, Takahashi A, Hanamata S, Ohno R, Hayashi T, Okada K, Koga J, Hirochika H, Yamane H, Kuchitsu K (2010) Regulation of microbe-associated molecular pattern-induced hypersensitive cell death, phytoalexin production, and defense gene expression by calcineurin B-like protein-interacting protein kinases, OsCIPK14/15, in rice cultured cells. Plant Physiol 153:678–692
9. Marino D, Dunand C, Puppo A, Pauly N (2012) A burst of plant NADPH oxidases. Trends Plant Sci 17:9–15

10. Suzuki N, Miller G, Morales J, Shulaev V, Torres MA, Mittler R (2011) Respiratory burst oxidases: the engines of ROS signaling. Curr Opin Plant Biol 14:691–699
11. Tena G, Boudsocq M, Sheen J (2011) Protein kinase signaling networks in plant innate immunity. Curr Opin Plant Biol 14:519–529
12. de la Torre F, Gutierrez-Beltran E, Pareja-Jaime Y, Chakravarthy S, Martin GB, Del Pozo O (2013) The tomato calcium sensor cbl10 and its interacting protein kinase CIPK6 define a signaling pathway in plant immunity. Plant Cell 25:2748–2764
13. Torres MA, Dangl JL (2005) Functions of the respiratory burst oxidase in biotic interactions, abiotic stress and development. Curr Opin Plant Biol 8:397–403
14. Torres MA, Dangl JL, Jones JD (2002) Arabidopsis gp91phox homologues AtrbohD and AtrbohF are required for accumulation of reactive oxygen intermediates in the plant defense response. Proc Natl Acad Sci USA 99:517–522

Chapter 10
Functional Role of CBL–CIPK in Plant Development

Abstract Extensive involvement of CBL–CIPKs in environmental stress pathways has been reported; at the same time, new reports are emerging, which suggests their participation in the plant developmental regulation. In this regard, calcium-mediated CBL–CIPK signaling pathway could possibly be acting as molecular link in regulating physiological and developmental processes in plants. With detail investigation of such connecting link, future strategies could be developed for reprogramming of developmental pathways influenced by environmental factors.

Keywords Function · Plant · Development · Pollen · Germination · Flower · Root · Seed · CBL · CIPK

10.1 Introduction

Much of the work on CBL–CIPKs is implicated in abiotic stress pathways in Arabidopsis and other plants species. However, a holistic genetic screening approach is required for investigating the role of CBL–CIPKs in plant growth and development. Recently, a few reports presented functional clues about specific CBL and CIPKs, which were involved in the developmental processes. In this section, we will be elaborating a few examples of CBLs and CIPKs involved in various developmental stages such as pollen germination and tube growth, flower, root, and seedling development.

10.2 Pollen Germination and Tube Growth

In a recent study by Mahs et al. [3], the role of CBL1 and CBL9 were explored in pollen germination and tube growth (Fig. 10.1). These *CBLs* are expressed in mature pollen and pollen tubes and targeting of these proteins to the plasma membrane is crucial for proper pollen germination and tube growth. Overexpression of

G.K. Pandey et al., *Global Comparative Analysis of CBL–CIPK Gene Families in Plants*, SpringerBriefs in Plant Science, DOI 10.1007/978-3-319-09078-8_10

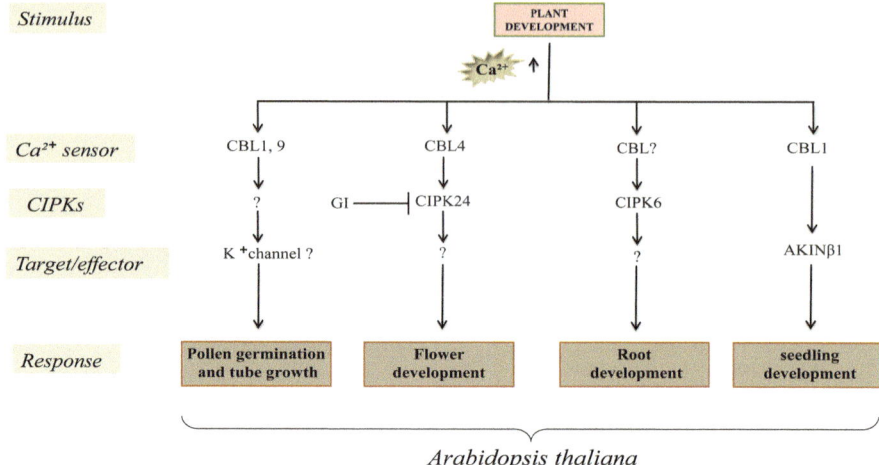

Fig. 10.1 Involvement of CBL–CIPK signaling pathways in plant development. This figure represents the four signaling pathways mediated by CBL–CIPK in Arabidopsis during various plant developmental processes

CBL1/CBL9 in Arabidopsis renders pollen germination and tube growth hypersensitive toward high external K^+ concentrations, while disruption of *CBL1* and *CBL9* reduces pollen tube growth under low K^+ conditions. The function of CBL1 and CBL9 is likely to be involved in the regulation of K^+ homeostasis in pollen germination and tube growth [3], whereas the CIPKs and K^+ channels involved in this process are still unknown.

10.3 Flower Development

The initiation of flowering in Arabidopsis is retarded or abolished by environmental stresses. A molecular explanation for this well-known fact can be explained from the following model. As per the model, GIGANTEA (GI) protein, a clock component important for flowering, act as a transitory regulator of SOS (salt overly sensitive) pathway activity whose presence or amount connects flowering to environmental conditions (Fig. 10.1) [1, 4]. GI formed complex with SOS2/CIPK24, because of that SOS2/CIPK24 is not able to activate SOS1 (Na^+/H^+-antiporter). This complex disintegrates in the presence of NaCl [4]. Moreover, rice SOS3 (OsCBL4) and OsSOS2 (OsCIPK24) exhibit a rhythmic and diurnal expression pattern [7]. This analysis establishes a cross-link between diurnal rhythm and SOS pathway in plants. In rice, one of the CIPK, i.e., *OsCIPK23,* is induced by multiple stresses and also regulates signaling pathways during pollination [9]. *Solanum lycopersicum, SlCIPK2,* is expressed specifically in the floral organ stamens [10]. SlCIPK2 interacts with SlCBLs and also with stress-responsive transcription factors such as SlERF7, SlCBF1, and SlAREB1 [10]. A well-defined function of flower-specific SlCIPK2 is still need to be investigated in tomato.

10.4 Root Development

One of the reports suggested role of CIPK6 in regulating the root development by affecting the shoot-to-root and root basipetal auxin transport [8]. The knockdown lines of *CIPK6* (RNAi lines *cipk6kd*) showed the pleotropic phenotypes such as fused cotyledons and swollen hypocotyl and compromised lateral root formation [8]. Based on the experimental facts, CIPK6 act as a positive regulator of root development and salt tolerance (Fig. 10.1) [8]. Similar functions were also shown by its ortholog in chickpea when ectopically expressed in tobacco [8]. Not only this, *SOS3/CBL4*, which expresses in the root at high level than shoot part, also modulates lateral root developmental plasticity and adaptation in response to low salt stress [11].

10.5 Seedling Development

In Arabidopsis, *CBL1* is induced by glucose and loss-of-function resulted in hypersensitivity to glucose and paclobutrazol (a GA biosynthetic inhibitor) [2]. Moreover, CBL1 protein physically interacts with AKINβ1, the regulatory β subunit of the SnRK1 complex, which has a central role in sugar signaling [2]. Identification of role of CBL1 in sugar signaling, seed germination, and seedling development has been linked with glucose and gibberellin responses (Fig. 10.1) [2]. Besides this, CIPK14 appeared to be regulated by both the circadian clock and PhyA, and involved in phytochrome A-mediated far-red light inhibition of greening in Arabidopsis seedlings [6]. In rice, OsCIPK31 has been identified to modulate responses to abiotic stresses during seed germination and seedling growth [5].

References

1. Kim WY, Ali Z, Park HJ, Park SJ, Cha JY, Perez-Hormaeche J, Quintero FJ, Shin G, Kim MR, Qiang Z, Ning L, Park HC, Lee SY, Bressan RA, Pardo JM, Bohnert HJ, Yun DJ (2013) Release of SOS2 kinase from sequestration with GIGANTEA determines salt tolerance in Arabidopsis. Nat Commun 4:1352
2. Li ZY, Xu ZS, Chen Y, He GY, Yang GX, Chen M, Li LC, Ma YZ (2013) A novel role for Arabidopsis CBL1 in affecting plant responses to glucose and gibberellin during germination and seedling development. PLoS ONE 8:e56412
3. Mahs A, Steinhorst L, Han JP, Shen LK, Wang Y, Kudla J (2013) The calcineurin B-like Ca^{2+} sensors CBL1 and CBL9 function in pollen germination and pollen tube growth in Arabidopsis. Mol Plant 6:1149–1162

4. Park HJ, Kim WY, Yun DJ (2013) A role for GIGANTEA: keeping the balance between flowering and salinity stress tolerance. Plant Signal Behav 8:e24820
5. Piao HL, Xuan YH, Park SH, Je BI, Park SJ, Kim CM, Huang J, Wang GK, Kim MJ, Kang SM, Lee IJ, Kwon TR, Kim YH, Yeo US, Yi G, Son D, Han CD (2010) OsCIPK31, a CBL-interacting protein kinase is involved in germination and seedling growth under abiotic stress conditions in rice plants. Mol Cells 30:19–27
6. Qin Y, Guo M, Li X, Xiong X, He C, Nie X, Liu X (2010) Stress responsive gene CIPK14 is involved in phytochrome A-mediated far-red light inhibition of greening in Arabidopsis. Sci China Life Sci 53:1307–1314
7. Soni P, Kumar G, Soda N, Singla-Pareek SL, Pareek A (2013) Salt overly sensitive pathway members are influenced by diurnal rhythm in rice. Plant Signal Behav 8:e24738
8. Tripathi V, Parasuraman B, Laxmi A, Chattopadhyay D (2009) CIPK6, a CBL-interacting protein kinase is required for development and salt tolerance in plants. Plant J 58:778–790
9. Yang W, Kong Z, Omo-Ikerodah E, Xu W, Li Q, Xue Y (2008) Calcineurin B-like interacting protein kinase OsCIPK23 functions in pollination and drought stress responses in rice (*Oryza sativa* L.). J Genet Genomics 35(531–543):s531–s532
10. Yuasa T, Ishibashi Y, Iwaya-Inoue M (2012) A flower specific calcineurin B-like molecule (CBL)-interacting protein kinase (CIPK) homolog in tomato cultivar micro-tom (*Solanum lycopersicum* L.). AJPS 3:753–763
11. Zhao Y, Wang T, Zhang W, Li X (2011) SOS3 mediates lateral root development under low salt stress through regulation of auxin redistribution and maxima in Arabidopsis. New Phytol 189:1122–1134

Chapter 11
Application and Future Perspectives of the CBL–CIPK Signaling

Abstract Climate change because of man-made activities is one of the major concerns of the world. Environmental stresses are the major outcomes of climate change, which adversely affect the crop productivity worldwide. On the top of it, increasing world population also poses a major threat to the natural resources and agriculture. Decreasing crop production due to various biotic and abiotic stresses is main concern of the time. Many of the calcium signaling components have been shown to be involved in regulating stress signaling. And exploring wide spread function of the calcium–CBL–CIPK signaling components will not only enable in-depth understanding of molecular mechanisms behind plant stress responses but could also be utilized in combating crop loss mediated by various stresses.

Keywords Applications · CBL–CIPK signaling · Questions

11.1 Basic Study Done So Far

Calcium ion is involved in diverse physiological and developmental processes. One of the important roles of calcium is to work as a signaling messenger, which regulates signal transduction in plants (calcineurin B-like protein). CBL is one of the calcium sensors that specifically interact with a family of serine–threonine protein kinases designated as CBL-interacting protein kinases (CIPKs). CBL and CIPK are two multi-gene families, which work in concert in the calcium signaling processes. The coordination of these two gene families defines complexity of the signaling networks in several stimulus-response coupling during various environmental stresses. From various studies on CBL and CIPK in Arabidopsis, this can be hypothesized that CBL–CIPK are emerging as key components of novel calcium signaling pathway in response to the different environmental stress conditions. Evolutionary studies revealed their existence in the single-cell algae to complex multi-cellular organisms. Phosphorylation not only regulates various downstream components in the CBL–CIPK signaling but also responsible for the stability of the CBL–CIPK complex. A few of the physiological targets of

© The Author(s) 2014
G.K. Pandey et al., *Global Comparative Analysis of CBL–CIPK Gene Families in Plants*, SpringerBriefs in Plant Science, DOI 10.1007/978-3-319-09078-8_11

CBL–CIPKs have been identified as transporters/channels and transcription factors regulating abiotic stress and nutrient signaling responses in Arabidopsis.

Several recent studies indicate that the function of the CBL–CIPK network is not restricted to Arabidopsis. The diverse spectrum of plants in which CBL and CIPK function was analyzed encompasses species such as rice, rape, wheat, apple, cotton, and poplar [2, 4, 5, 14, 18, 20–22]. Comparative structural analyses of CBLs and CIPKs from Arabidopsis, pea and rice revealed a high degree of conservation and in conjunction with this functional analyses also suggested comparable mechanisms underlying signal transduction via CBL–CIPK complexes from these species [12, 17].

11.2   Applications of the CBL–CIPK Signaling System

Because of increasing climate change and adverse environmental conditions, there is a decrease in the yield of various food crops, which presents an alarming situation for the world food security. This enforces to develop varieties, which are tolerant to multiple stresses. To attain this, in-depth understanding of abiotic stress-mediated signaling and the key components involved in regulating these responses are required. Many signaling pathways regulate the stress-mediated signaling, and calcium is one of the pivotal signaling molecules majorly responsible for controlling diverse arrays of abiotic and biotic stresses. In the calcium signaling pathway, CBL–CIPK module is extensively studied to be involved in regulating stress responses in Arabidopsis. With usage of genetic, biochemical, cell biological, and physiological approaches, molecular mechanism of stress signal transduction has been identified for several CBL–CIPK modules. With holistic understanding of mechanism of stress signaling and adapting responses in Arabidopsis, which is a model plant, this acquired knowledge can be translated to crop plants for the development of varieties tolerant to multiple abiotic stresses and enhancing the quantity and quality of the crop production.

Although not many of the CBL–CIPK components have been functionally characterized in other plant species, their identification and involvement in abiotic stress signaling pathway is being reported. Out of 30 OsCIPKs, 20 of them respond to at least one of abiotic stresses such as drought, salt, low temperature, and ABA in rice [19]. Overexpression of *OsCIPK03*, *OsCIPK12,* and *OsCIPK15* leads tolerance to cold, drought, and salt stress, respectively, in rice. Orthologous genes are related gene between species, and exploring the function of orthologous genes in crop plants has also led to some lead in identification of functional similarities, one of such example is conservation of SOS pathway among Arabidopsis and other crop plants.

The clue of orthologous CBL–CIPK signaling components in different species with high sequence similarities suggested conservation of functions performed by these in the signaling pathway. In terms of target regulation by CBL–CIPK module, it seems that research from Arabidopsis can also be extrapolated to other plant species, for example, the regulation of potassium channels. The shaker potassium channel VvK1.1 from grape, which is the counterpart of AtAKT1, could be stimulated by co-expression of AtCIPK23–AtCBL1 in *Xenopus* oocytes [3]. Moreover, the SOS pathway for salt tolerance that was originally identified in Arabidopsis is also conserved in rice and other species at the target level [7–10, 13, 16].

To understand the mechanism of functional basis of CBL–CIPK network, it is very important to study lower organisms in plant kingdom such as algae and mosses, which have less complex CBL–CIPK networks because of lesser number of *CBL–CIPK* genes. In addition, there are some recent reports emerging for the presence and functional role of CBL–CIPK system in organisms like halophyte (e.g., *Hordeum brevisubulatum*) and xerophyte (e.g., *Ammopiptanthus mongolicus*), which will unearth new layers of information for the diverse and holistic functioning of CBL–CIPK components in stress-mediated signaling pathways [1, 6, 11, 15].

11.3 Future of CBL–CIPK Signaling

It is evident that every discovery has much to explore in future. Based on the ability to form CBL–CIPK pair, these proteins must exist as a large part of calcium-regulated interactome inside the cell. It is quite imperative to understand the specificity and cross talk in signaling pathways mediated by this large possibility of CBL–CIPK interactome under different conditions, to map the definite signaling pathways. With the discovery of CBL and CIPK components in lower to higher plants, the identification of various targets of the CBL–CIPK still required extensive research.

Not only in the regulation of abiotic stresses the role of CBL–CIPK signaling pathway in other aspects such as plant development and biotic stress have also pave the way for in-depth future work to understand the overlap and cross talk in signaling pathways during physiological as well as developmental responses. In future, a greater attention is required to understand cross talk and interconnection of CBL–CIPK signaling with other signaling pathway such as MAP kinases, CDPK (calcium dependent protein kinases), phytohormone such as ABA (abscisic acid), (jasmonic acid) JA, SA (salicylic acid), ethylene, and auxin-mediated signaling network.

Based on the previous research in calcium-mediated signaling pathway in Arabidopsis, a few genetic components such as CBL–CIPK and their targets have been identified, which could be the key player in genetically manipulating the crop to achieve higher yield and productivity under abiotic stress conditions. At the same time, a detailed molecular understanding of these calcium-mediated signaling cascades need to be investigated in detail, which will further enhance our understanding how these abiotic stress signaling mechanisms operate at whole plant physiological level. A holistic exploration of calcium-mediated CBL–CIPK pathway in economically important crops is very necessary to develop newer varieties for agricultural enhancement, which can sustain higher degree of environmental stress without affecting the yield and productivity under multiple abiotic stress conditions.

11.4 Questions for the Future

In addition to future perspectives, there are few key questions listed below, which will definitely enable plant biologist to develop a better and holistic understanding of stress signal transductions pathways in plants.

1. Because of large number of genes for both *CBLs* and *CIPKs*, it is obvious to ascertain how many and which *CBLs* and *CIPKs* provide stress responses, which genes make essential contributions to stress tolerance, and how many and which mechanisms serve in stress protection signaling?
2. Which *CBLs* and *CIPKs* are confined to one stress signaling pathways or interconnected with other signaling pathways?
3. What are the specific substrates of CBL–CIPK module, and which particular protein phosphatases act upon to counteract the function of a particular CIPK in a respective CBL–CIPK signaling pathway?
4. What are the downstream signaling components regulated by only CBL in the calcium signaling pathways?
5. What is the mechanistic action of a particular CBL–CIPK in a designated signaling pathway? Is it only phosphorylation based modulation of downstream target or any other cellular regulation involved under a defined condition?
6. How specificity and overlap is established in CBL–CIPK interactome and how exactly this pathway is interconnected with other signaling pathway and their component in the cell?
7. How CBL–CIPK pathway is linked to the upstream and downstream component of calcium signature or transient increase in cytosolic Ca^{2+}, especially with perception of signal and calcium homeostasis machinery in the cell?

References

1. Chen JH, Sun Y, Sun F, Xia XL, Yin WL (2011) Tobacco plants ectopically expressing the *Ammopiptanthus mongolicus* AmCBL1 gene display enhanced tolerance to multiple abiotic stresses. Plant Growth Regul 63:259–269
2. Chen L, Ren F, Zhou L, Wang QQ, Zhong H, Li XB (2012) The *Brassica napus* calcineurin B-Like 1/CBL-interacting protein kinase 6 (CBL1/CIPK6) component is involved in the plant response to abiotic stress and ABA signalling. J Exp Bot 63:6211–6222
3. Cuellar T, Pascaud F, Verdeil JL, Torregrosa L, Adam-Blondon AF, Thibaud JB, Sentenac H, Gaillard I (2010) A grapevine Shaker inward K^+ channel activated by the calcineurin B-like calcium sensor 1-protein kinase CIPK23 network is expressed in grape berries under drought stress conditions. Plant J 61:58–69
4. Deng X, Hu W, Wei S, Zhou S, Zhang F, Han J, Chen L, Li Y, Feng J, Fang B, Luo Q, Li S, Liu Y, Yang G, He G (2013) TaCIPK29, a CBL-interacting protein kinase gene from wheat, confers salt stress tolerance in transgenic tobacco. PLoS One 8:e69881
5. Gao P, Zhao PM, Wang J, Wang HY, Du XM, Wang GL, Xia GX (2008) Co-expression and preferential interaction between two calcineurin B-like proteins and a CBL-interacting protein kinase from cotton. Plant Physiol Biochem 46:935–940

6. Guo L, Yu Y, Xia X, Yin W (2010) Identification and functional characterisation of the promoter of the calcium sensor gene CBL1 from the xerophyte *Ammopiptanthus mongolicus*. BMC Plant Biol. 10:18:doi: 10.1186/1471-2229-10-18

7. Hu DG, Li M, Luo H, Dong QL, Yao YX, You CX, Hao YJ (2012) Molecular cloning and functional characterization of MdSOS2 reveals its involvement in salt tolerance in apple callus and Arabidopsis. Plant Cell Rep 31:713–722

8. Huertas R, Olias R, Eljakaoui Z, Galvez FJ, Li J, De Morales PA, Belver A, Rodriguez-Rosales MP (2012) Overexpression of SlSOS2 (SlCIPK24) confers salt tolerance to transgenic tomato. Plant Cell Environ 35:82–1467

9. Ji H, Pardo JM, Batelli G, Van Oosten MJ, Bressan RA, Li X (2013) The Salt Overly Sensitive (SOS) pathway: established and emerging roles. Mol Plant 6:86–275

10. Kanwar P, Sanyal SK, Tokas I, Yadav AK, Pandey A, Kapoor S, Pandey GK (2014) Comprehensive structural, interaction and expression analysis of CBL and CIPK complement during abiotic stresses and development in rice. Cell calcium (in press)

11. Li R1, Zhang J, Wu G, Wang H, Chen Y, Wei J. (2012) HbCIPK2, a novel CBL-interacting protein kinase from halophyte Hordeum brevisubulatum, confers salt and osmotic stress tolerance. Plant Cell Environ. 2012 Sep;35(9):1582-600. doi: 10.1111/j.1365-3040.2012.02511.x. Epub 2012 Apr 27

12. Mahajan S, Sopory SK, Tuteja N (2006) Cloning and characterization of CBL–CIPK signalling components from a legume (*Pisum sativum*). FEBS J 273:25–907

13. Martinez-Atienza J, Jiang X, Garciadeblas B, Mendoza I, Zhu JK, Pardo JM, Quintero FJ (2007) Conservation of the salt overly sensitive pathway in rice. Plant Physiol 143:1001–1012

14. Piao HL, Xuan YH, Park SH, Je BI, Park SJ, Kim CM, Huang J, Wang GK, Kim MJ, Kang SM, Lee IJ, Kwon TR, Kim YH, Yeo US, Yi G, Son D, Han CD (2010) OsCIPK31, a CBL-interacting protein kinase is involved in germination and seedling growth under abiotic stress conditions in rice plants. Mol Cells 30:19–27

15. Shang G, Cang H, Liu Z, Gao W, Bi R (2010) Crystallization and preliminary crystallographic analysis of a calcineurin B-like protein 1 (CBL1) mutant from *Ammopiptanthus mongolicus*. Acta Crystallogr Sect F Struct Biol Cryst Commun 66:1602–1605

16. Tang RJ, Liu H, Bao Y, Lv QD, Yang L, Zhang HX (2010) The woody plant poplar has a functionally conserved salt overly sensitive pathway in response to salinity stress. Plant Mol Biol 74:367–380

17. Tuteja N, Mahajan S (2007) Further characterization of calcineurin B-like protein and its interacting partner CBL-interacting protein kinase from pisum sativum. Plant Signal Behav 2:358–361

18. Wang RK, Li LL, Cao ZH, Zhao Q, Li M, Zhang LY, Hao YJ (2012) Molecular cloning and functional characterization of a novel apple MdCIPK6L gene reveals its involvement in multiple abiotic stress tolerance in transgenic plants. Plant Mol Biol 79:123–135

19. Xiang Y, Huang Y, Xiong L (2007) Characterization of stress-responsive CIPK genes in rice for stress tolerance improvement. Plant Physiol 144:1416–1428

20. Yang W, Kong Z, Omo-Ikerodah E, Xu W, Li Q, Xue Y (2008) Calcineurin B-like interacting protein kinase OsCIPK23 functions in pollination and drought stress responses in rice (*Oryza sativa L.*). J Genet Genomics 35(531–543):s1–s2

21. Zhang H, Lv F, Han X, Xia X, Yin W (2013) The calcium sensor PeCBL1, interacting with PeCIPK24/25 and PeCIPK26, regulates Na^+/K^+ homeostasis in *Populus euphratica*. Plant Cell Rep 32:611–621

22. Zhang H, Yin W, Xia X (2010) Shaker-like potassium channels in Populus, regulated by the CBL–CIPK signal transduction pathway, increase tolerance to low-K^+ stress. Plant Cell Rep 29:1007–1012